Springer Theses

Recognizing Outstanding Ph.D. Research

For further volumes:
http://www.springer.com/series/8790

Aims and Scope

The series "Springer Theses" brings together a selection of the very best Ph.D. theses from around the world and across the physical sciences. Nominated and endorsed by two recognized specialists, each published volume has been selected for its scientific excellence and the high impact of its contents for the pertinent field of research. For greater accessibility to non-specialists, the published versions include an extended introduction, as well as a foreword by the student's supervisor explaining the special relevance of the work for the field. As a whole, the series will provide a valuable resource both for newcomers to the research fields described, and for other scientists seeking detailed background information on special questions. Finally, it provides an accredited documentation of the valuable contributions made by today's younger generation of scientists.

Theses are accepted into the series by invited nomination only and must fulfill all of the following criteria

- They must be written in good English.
- The topic should fall within the confines of Chemistry, Physics, Earth Sciences, Engineering and related interdisciplinary fields such as Materials, Nanoscience, Chemical Engineering, Complex Systems and Biophysics.
- The work reported in the thesis must represent a significant scientific advance.
- If the thesis includes previously published material, permission to reproduce this must be gained from the respective copyright holder.
- They must have been examined and passed during the 12 months prior to nomination.
- Each thesis should include a foreword by the supervisor outlining the significance of its content.
- The theses should have a clearly defined structure including an introduction accessible to scientists not expert in that particular field.

Oleg Pronin

Towards a Compact Thin-Disk-Based Femtosecond XUV Source

Doctoral Thesis accepted by
Ludwig-Maximilians-University Munich, Germany

Springer

Author
Dr. Oleg Pronin
Department of Physics
Ludwig-Maximilians-University Munich
Garching
Germany

Supervisor
Prof. Dr. Ferenc Krausz
Laboratory of Attosecond physics
Max-Planck Institute for Quantum Optics
Garching
Germany

ISSN 2190-5053 ISSN 2190-5061 (electronic)
ISBN 978-3-319-01510-1 ISBN 978-3-319-01511-8 (eBook)
DOI 10.1007/978-3-319-01511-8
Springer Cham Heidelberg New York Dordrecht London

Library of Congress Control Number: 2013944534

Printed on acid-free paper

Springer is part of Springer Science+Business Media (www.springer.com)

To my teacher Igor P. Simonov

Supervisor's Foreword

Science and technology need femtosecond (fs) laser oscillators of a much higher average power than what the market currently offers. Such lasers are indispensable for material modifications and for seeding different laser sources, including parametric amplifiers, regenerative amplifiers, and enhancement cavities. A very promising candidate for a new generation of such oscillators is the Yb-based thin-disk laser oscillator. In the foreseeable future, this laser will replace the now 30-year-old Ti:Sa femtosecond laser, a workhorse in hundreds of research labs around the world. To make this happen, the breakthrough to sub-10 phase-stabilized pulses has to be achieved, where the phase means the time shift of the maximum of the highest electromagnetic oscillation relative to the maximum of the virtual pulse envelope. Few-cycle pulses are necessary for dozens of time-resolved experiments with a correspondingly broad spectrum for broadband frequency comb experiments. The main parameter that makes an Yb-based oscillator a strong candidate for the next generation of femtosecond lasers is its average power, which is a factor of 10–100 higher than that of Ti:Sa oscillators.

Readers may be surprised to learn that the laser oscillator has power limitations. This is mainly because of thermal effects in the active medium. It is worth noting here that the average power of the thin-disk laser is the highest among other types of lasers, owing to the geometry of the active medium. A thin active medium can—thanks to its efficient cooling geometry—tolerate kW-level light power without any parasitic effects.

The first task assigned to this Ph.D. project was finding a mode-locking mechanism faster than the well-established technique based on the saturable Bragg reflector. The known technique provided only 1 ps pulses which is much longer than what is delivered by the Ti:Sa laser based on the Kerr-lens mode-locking technique. The Kerr-lens mode-locking is the fastest mechanism known so far. Oleg Pronin successfully implemented this mechanism in his laser oscillator, which resulted in pulses down to 200 fs. The second task was pushing the pulse duration even further down—to sub-50 fs. This was done with photonic crystal fibers and a chirped (dispersive) mirror compressor. So far, all known previous attempts at a similar average power of tens of watts have led to fiber damage. By optimizing all the critical parameters, Oleg Pronin demonstrated pulses down to 40 fs by using only one stage of this broadening/compression scheme. The third task was connected to the stabilization of the carrier-envelope phase, as it is called, of this short pulse. Traditional phase stabilization approaches were not capable of fixing the phase of the Yb oscillators pulses with its longer pulses, higher average power, and less passive stability (higher timing jitter) in comparison to the schemes of other laser types. The approach Oleg Pronin proposed—and the first results of which look very promising—is based on the fast modulation of the diode pump parameters, giving reason to hope that the goal of this task will be achieved soon. As an interesting application of the laser oscillator developed in his Ph.D. work, Oleg Pronin has presented first steps toward intracavity XUV generation: pulse energy limitations due to instabilities, and a very broadband XUV output coupler.

Besides reporting these original results of research, this thesis also provides a broad introduction to the physics of modern mode-locking techniques and to the thin-disk laser concept. The results described here and illustrated in 69 figures with 202 references enable the reader to obtain an even deeper insight into this exciting field of research.

Munich, June 2013 Prof. Dr. Ferenc Krausz

Acknowledgments

First of all, I would like to thank Prof. Ferenc Krausz for giving me the opportunity to work on extremely interesting topics in the excellent working environment with excellent and motivated people.

I wish to thank Alexander Apolonskiy for his support and for giving me that large degree of independence in my research work and always believing in a successful outcome of the experiment.

Many thanks to the very skilled experimentalist Catherine Teisset who took me under her wing during the beginning of my work and for her patience, continuous explanations and support.

I am indebted to Vladimir Pervak for helping me with adapting to Germany and, of course, for designing and manufacturing different types of coatings and dispersive mirrors, for always having time for discussions, and providing supports for the different coating characterization apparatus. I am thankful to his team members Ivan Angelov, Olga Razskazovskaya, and Michael Trubetskov, who helped me with multilayer designs and their characterization.

I am grateful to Vladimir Kalashnikov for that large amount of time he spent with me explaining and discussing (mostly via skype) the theoretical models of mode-locking, for his interest in KLM thin-disk oscillators, and for many simulations shedding the light on the experimental work.

Many thanks to the team from Walter Schottky Institute, to Prof. Marcus-Christian Amann, and Christian Grasse, for the fruitful collaboration in SESAM manufacturing.

Christian, this was the fastest and most efficient collaboration I ever experienced.

Special thanks go to Fabian Lücking, who started the CEP stabilization experiment and is always open to share his knowledge and expertise. The efficiency of those months working together have been astonishing to me.

I would like to thank Jonathan Brons for his dedicated work in the lab and always being helpful. Obviously, without you that progress with the KLM TD oscillator, or any oscillator at all, would not be possible. Here, I want to thank once more Alexander Apolonskiy, Jonathan Brons, and Vladimir Kalashnikov for reading this thesis and making their invaluable comments.

I thank Jens Rauschenberger and Ioachim Pupeza for familiarizing me with enhancement cavities and the challenges of coupling out XUV light.

I appreciate the fruitful discussions and teaching of Thomas Udem and his direct help in the lab.

I would like to thank Mikhail Larionov for many very useful technical and physical recommendations and replies regarding the thin-disk technology.

I would like to thank Prof. Thomas Dekorsy, Dirk Sutter, Dominik Bauer, and Farina Schättiger for providing the help with SESAM characterization and sharing their knowledge and opinions.

This work would not be possible without our Werkstatt and Mr. Öhm. I am grateful not only for the many many components and installations they made but also for the fact that the latter was made with no drawings from my side. Thanks a lot to Mr. Herbst for the support with electronics and drivers.

I appreciate the help and discussions with Ernst Fill, Thomas Metzger, Hanieh Fattahi, Alexander Schwarz, Roswitha Graf, Moritz Ueffing, Zsuzsanna Major, Thomas Ganz, Friedrich Kirchner, Alexander Gliserin, Henning Carstens, Simon Holzberger, Yunpei Deng, Sergiu Amarie, and Vizbaras Kristijonas.

I wish to thank Matthias Kling who kindly directed me to the IMPRS when I was searching for a Ph.D. position at MPQ.

Special thanks to Mrs. Wild who organized countless things for me and without whom my arrival and life in Germany could not have been as pleasant as it is. Many thanks to Katharina Adler, Martin Groß, Franziska Hoss, Valentina Rot, and Klaus Franke for taking care of the numerous bureaucratic and organizational challenges.

I wish to thank all members in our group, who were not mentioned before, for the nice and friendly environment.

Finally, I would like to thank my family and Maria for their constant support.

Abstract

The goal of this thesis is to develop a compact high-power solid-state oscillator capable of superseding existing ultrafast technology based on low-power Ti:sapphire oscillators. Different applications such as extra- or intra-cavity XUV generation, seeding of high-energy low-repetition-rate amplifier systems and femtosecond enhancement cavities can be dramatically influenced by the availability of such a reliable, compact femtosecond source. We applied, for the first time, Kerr-lens mode-locking to a thin-disk Yb:YAG oscillator, resulting in an unprecedented combination of an average power 45 W and pulse duration of 250 fs directly available from the oscillator with repetition rate of 40 MHz and a footprint of only 1×0.4 m^2. Even shorter emission-bandwidth-limited 200 fs pulses have been generated with the reduced output coupler transmission of 5.5 % at an average power of 17 W. Moreover, the oscillator was operating not only in the negative dispersion regime common to solid-state oscillators but also in the positive dispersion regime, resulting in a spectrum spanning a range of 20 nm, which is the broadest hitherto reported for Yb:YAG material in high-power operation.

First attempts toward CE phase-stabilized high-power pulses from such an oscillator are also described.

State-of-the-art XUV generation driven by high-power NIR femtosecond systems requires methods of separating generated XUV from NIR radiation. Such a method has been proposed and realized. It constitutes a glass substrate having a low-loss anti-reflection coating for NIR wavelengths at grazing incidence of >70° and serving simultaneously as a high reflector for radiation in the range of 1–100 nm with reflectivity >60 %. The device can be used for both extra-and intra-cavity XUV generation.

Contents

Acronyms

A-FPSA	Antiresonant Fabry Pero saturable absorber
AOM	Acousto-optical modulator
APD	Avalanche photodiode
CE	Carrier-envelope
CPA	Chirped-pulse amplification
CW	Continuous wave
DF	Driving field
EOM	Electro-optical modulator
FWHM	Full width at half maximum
GDD	Group delay dispersion
GIP	Grazing-incidence plate
HD	High-dispersive
HHG	High-harmonic generation
HR	High reflector
IR	Infrared
ISA	Inverse saturable absorption
KLM	Kerr-lens mode-locking
LT	Low temperature
MBE	Molecular Beam Epitaxy
ML	Mode-locked
MOCVD	Molecular Chemical Vapor Deposition
MQW	Multiple quantum well
NIR	Near-infrared radiation
NOPA	Non-collinear optical parametric amplification
NLS	Nonlinear Schrödinger equation
PCF	Photonic crystal fibre
PLL	Phase-locked loop
PPLN	Periodically poled lithium niobate
QML	Q-switched mode-locking
QW	Quantum well
ROC	Radius of curvature
SA	Saturable absorber
SAM	Self-amplitude modulation
SESAM	Semiconductor Saturable Absorber Mirror

SPM	Self-phase modulation
SNR	Signal to noise ratio
TD	Thin disk
Ti:Sa	Ti:sapphire
TPA	Two-photon absorption
VBG	Volume bragg grating
XUV	Extreme ultraviolet radiation

Chapter 1
Introduction

The invention of the optical maser [1–3] has had an enormous impact on our life and technology. The laser made data storage (CDs) and the internet possible, and became a golden standard in eye surgery, cancer treatment and diagnostics, brought unprecedented precision and speeds in micromachining, paved the way for clean energy sources and more. Even after more than 50 years it keeps developing, conquering and opening new applications. Laser physics, quantum electronics and photonics are fast expanding scientific areas to which the laser gave birth.

Just to further illustrate this enormous progress: the invention of the first mode-locked He-Ne laser in 1964 [4] with 2.5-ns pulse duration initiated developments which culminated in <10-fs pulses (a few light cycles). This evolution took an intricate way: from the gas laser to the solid-state laser [5], then to the liquid dye laser [6] and more recently to the all-solid-state diode-pumped Ti:sapphire oscillator [7, 8]. The latter technology being robust, compact and durable breathed new life into already exciting applications and opened up fundamentally new directions in photon science. Ti:sapphire ultrafast technology is also not the last step in this evolution and Yb thin-disk technology may very likely be on the way to superseding it.

1.1 Motivation

Femtosecond lasers contributed to the observation of intermediate states of chemical reactions, thus making femtochemistry possible [9]. They enabled high-precision frequency comb spectroscopy [10, 11] and even shorter attosecond pulses with durations down to 80 as [12, 13] could be generated. The latter gave birth to attosecond physics combining high-field phenomena, extreme ultraviolet and X-ray ranges with a never before seen ability to resolve electron dynamics and interactions on its attosecond time scale [14]. The most common way of attosecond pulse generation is high-harmonic generation (HHG) in a noble gas. For efficient generation of single attosecond pulses and utilizing phase matching one needs few-cycle pulses. The

O. Pronin, *Towards a Compact Thin-Disk-Based Femtosecond XUV Source*,
Springer Theses, DOI: 10.1007/978-3-319-01511-8_1,

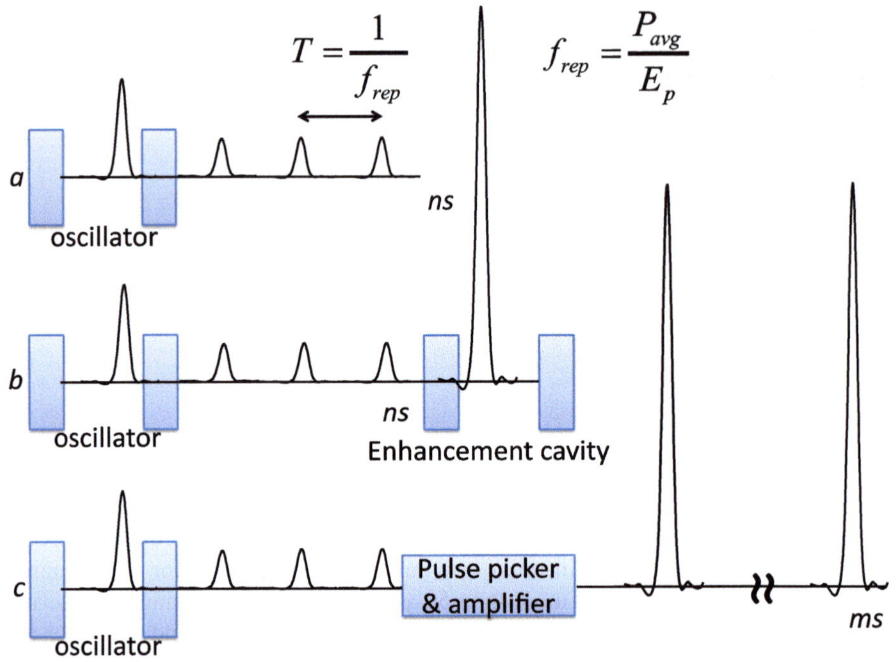

Fig. 1.1 Schematic of different approaches to drive high-field experiments. **a** directly by oscillator, **b** inside the enhancement cavity, **c** by CPA system

latter, however, is not a necessary condition for XUV generation and longer pulse durations can be employed. In each case high intensities exceeding 10^{13} W/cm^2 on the gas target are necessary. The routine method for generating high harmonics is schematically shown in Fig. 1.1c. The pulses coming from the seed oscillator have to be picked up at somewhat lower repetition rate (a few kHz), stretched, amplified and compressed. The procedure is indeed even more complex, involving NOPA and hollow-core fibres for spectral broadening and consequent compression. The simple relation

$$E_p = \frac{P_{avg}}{f_{rep}},$$

where E_p is the pulse energy, P_{avg} the average power and f_{rep} the repetition rate, indicates that due to the low repetition rate even at a moderate power of several watts several mJ can be reached. Due to the very low conversion efficiency (10^{-8}–10^{-6}) of the HHG process and the low average power amplifier systems driving at kHz repetition rate, the resulting average XUV power is well below the mW level. This fact restricts the field of HHG applications. High-resolution spectroscopy with XUV frequency combs [15, 16], pump-probe measurements, photoelectron emission microscopy, photoelectron imaging spectroscopy, pump-probe diffraction experiments with electrons and nanostructure characterization could substantially benefit from MHz-repetition-rate XUV sources. For many of these experiments the output of the signal level is very low and data acquisition over many hours or days is necessary when operated at ~kHz repetition rate. A difference of three to four orders can be gained by working at MHz repetition rate. Experiments such as electron diffraction with fs time resolution can be enabled only in the single electron regime (because of the space charge) and ultimately need MHz laser sources.

Another approach seen in Fig. 1.1b is to use a passive femtosecond enhancement resonator for storage of radiation. By using this method, one can increase the power from a mode-locked seed oscillator by a large factor (10–10^4) inside the cavity. Extremely low-loss optics and fine dispersion control of the cavity are needed. Moreover, the cavity has to be kept in resonance with the seeding oscillator and mode-matched to it. Nevertheless, due to the low XUV conversion efficiency an enhancement cavity recycles circulating pulses with rather small influence on the enhancement factor and remains the most promising tool in accessing HHG at MHz repetition rate.

The idea of utilizing the high average powers inside a laser oscillator cavity has become a reality with recent progress in high-power femtosecond thin-disk lasers. Typically, the output coupler transmission T of an oscillator amounts to only a few per cent. This means that power stored inside the oscillator cavity is a factor $\sim 1/T$ higher than the output power (see Fig. 1.1c). On the other hand, the oscillator is a dynamic nonlinear system which cannot simply be operated at arbitrarily high energies due to the instabilities of mode-locking.

As can be seen, there is no unique and universal laser tool for HHG generation. All schemes have their prospective application fields and niches. Yet for any of the schemes in Fig. 1.1 the oscillator continues to be the decisive element. An energy-upscaled seed oscillator (>μJ) can eliminate the need for pre-amplification stages and thus simplify the system. A lesser gain needed from the amplifier results in less gain narrowing and shorter pulses after compression can be realized. The enhancement cavity combined with a high-energy table-top compact oscillator becomes an attractive tool for XUV generation in many labs. The constraints on the enhancement factor can be substantially lowered when seed oscillators with higher pulse energies become available. Progress on the development of a Kerr-lens thin-disk oscillator allows intra-cavity peak intensities similar to those of the first generation of enhancement cavities [15].

Other decisive elements for XUV generation are an XUV/IR beam splitter and XUV output coupler. With recent progress in power scaling of femtosecond high-power oscillators [17, 18] and amplifiers [19, 20] XUV/IR beam splitters able to withstand high average power and peak intensity are necessary. Also the generation of XUV inside the oscillator or enhancement cavity (Fig. 1.1a, b) demands that the element be able to outcouple XUV light from resonator. Such an element and method have been proposed and realized in this work.

Indeed the application area of thin-disk femtosecond oscillators is not restricted to only XUV generation—many other wavelength ranges can be reached as well. THz radiation, UV and VUV generation is possible by means of ultrafast high power thin-disk oscillators, which in turn paves the way to other various and exciting applications.

1.2 State-of-the-Art TD Oscillators

Bulk oscillators are not power-scalable and their pulse-energies can only be increased by lowering the repetition rate. Thus the development of Ti:Sa oscillators culminated in pulse energies of >500 nJ [21] and more recently >650 nJ in commercial systems [22] with 50-fs pulse duration. Ti:Sa oscillators generating sub-10 fs [23] and even sub-6 fs [24] with several nJ pulse energies were developed. The average power of Ti:Sa bulk oscillators remains in the range of several watts. Other Yb bulk oscillators with different gain media could approach the level of 10 W [25]. Fibre and innoslab amplifier systems [19, 20] are operating in a way similar to that depicted in Fig. 1.1c, but at MHz repetition rate and corresponding power levels approaching 1 kW. These systems continue to be fairly complex, bulky and expensive. The power scalability of thin-disk technology makes it a promising candidate for developing the next generation of femtosecond oscillators, yielding average output powers exceeding 140 W [26, 27], pulse energies of 40 μJ [26] and pulse durations of 700–1000 fs. The progress in the development of thin-disk oscillators is depicted in Fig. 1.2. Until recently, all thin-disk oscillators had been mode-locked with semiconductor saturable absorber mirrors (SESAM). Their moderate modulation depth, and finite relaxation time prevented the generation of pulses shorter than 700 fs in Yb:YAG (see black dots in Fig. 1.2). In contrast, nearly gain-bandwidth-limited pulse generation from a thin-disk Yb:YAG oscillator is demonstrated via KLM in this work (red dots in Fig. 1.2). It will be shown further that pulses shorter than 200 fs can be generated even from Yb:YAG and even shorter in other Yb-doped thin-disk oscillators. The many shortcomings originating from SESAM are simply not intrinsic to the KLM method.

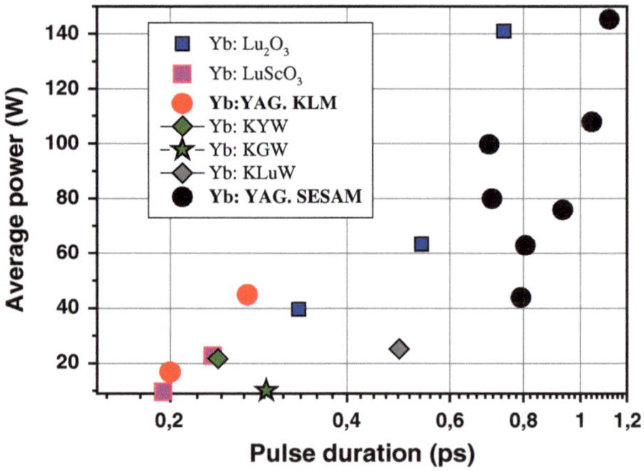

Fig. 1.2 State-of-the-art in the development of TD oscillators. *Red circles* correspond to the results presented in this work. The oscillators with power levels below 10 W were omitted because bulk oscillators with average power levels in this range were demonstrated [25], thus showing no need for the thin-disk geometry. References and details can be found in Table A.1.

1.3 Structure of the Thesis

This thesis is focused on the development of high-power/energy thin-disk femtosecond oscillator and the method for coupling out the XUV light which can be generated inside this oscillator.

The second chapter describes the basics of soliton mode-locking as well as different factors influencing pulse formation and propagation. The complex nonlinear Ginzburg-Landau equation is introduced here as well as its relation to the nonlinear Schrödinger equation. A qualitative explanation of soliton mode-locking in the regime of normal and anomalous dispersion is given. The assumptions of the theoretical models are described. This chapter serves as a theoretical basis for the experimentalist involved in oscillator development.

The third chapter introduces the thin-disk concept and its main features and advantages, particularly its power scalability. Different gain media are considered for thin-disk operation. Yb:YAG and Yb:Lu_2O_3 are discussed in more detail as the most prominent ones. The influence of a thermal lens on the resonator stability and its misalignment sensitivity as well as the possibility to increase considerably the mode areas over intracavity optical elements are reported.

Chapter 4 describes the experimental work on oscillator mode-locking with a SESAM. The SESAM parameters and structure are briefly introduced. Experiments with different SESAMs and cavity configurations are shown, thus forming a complete picture of the limitations and prospects of the SESAM approach.

Chapter 5 explains the basics of Kerr-lens mode-locking, the way to realize it in a thin-disk oscillator and problems involved. Different oscillator configurations and parameters, including the stability, are presented. Preliminary results on Kerr-lens mode-locking of Yb:YAG with positive GDD and Yb:Lu_2O_3 with negative GDD are shown. The prospects of energy scaling and pulse shortening as well as the advantages and disadvantages of Kerr-lens mode-locking and SESAM-mode-locking are summarized at the end of the chapter. Chapter 6 describes experiments on spectral broadening and temporal compression of the output of the KLM oscillator developed and CE phase stabilization of this oscillator. The spectral broadening was performed in photonic crystal fibre at full oscillator power. The pulses compressed afterwards were used for CE frequency detection. Stability of the CE phase was characterized and preliminary experiments on its locking were performed.

Chapter 7 deals with the problem of separating the XUV from the driving field and coupling the XUV light out of the resonator. A detailed overview of different XUV output coupling methods is presented. A novel method based on the idea of using grazing incidence is realized experimentally and characterized in the XUV and infrared spectral regions. The possibilities and prospects of using it in an enhancement cavity, inside a laser-oscillator cavity or after a laser amplifier are discussed.

In conclusion, the experimental results are summarized and further prospective developments are discussed.

All data presented in this work are stored on the data archive server of our group (see Appendix C).

References

1. N. Basov, A. Prokhorov, Application of molecular beams for the radiospectroscopic study of rotational molecular spectra. Sov. Phys. JETP **27**, 431–438 (1954)
2. J.P. Gordon, H.J. Zeiger, C.H. Townes, Molecular microwave oscillator and new hyperfine structure in the microwave spectrum of NH_3. Phys. Rev. **95**, 282–284 (1954)
3. T. Maiman, Stimulated optical radiation in ruby. Nature **187**, 493–494 (1960)
4. L.E. Hargrove, R.L. Fork, M.A. Pollack, Locking of He-Ne laser modes induced by synchronous intracavity modulation. Appl. Phys. Lett. **5**(1), 4–5 (1964)
5. A.J. DeMaria, W.H. Glenn, M.J. Brienza, M.E. Mack, Picosecond laser pulses. Proc. IEEE **57**(1), 2–25 (1969)
6. E. Ippen, C. Shank, A. Dienes, Passive mode locking of the cw dye laser. Appl. Phys. Lett. **21**(8), 348–350 (1972)
7. D.E. Spence, P.N. Kean, W. Sibbett, 60-fsec pulse generation from a self-mode-locked Ti:sapphire laser. Opt. Lett. **16**(1), 42–44 (1991)
8. F. Krausz, M. Fermann, T. Brabec, P. Curley, M. Hofer, M. Ober, C. Spielmann, E. Wintner, A. Schmidt, Femtosecond solid-state lasers. IEEE J. Quantum Electron. **28**(10), 2097–2122 (1992)
9. A.H. Zewail, Laser femtochemistry. Science **242**(4886), 1645–1653 (1988)
10. R. Holzwarth, T. Udem, T.W. Hänsch, J.C. Knight, W.J. Wadsworth, P.S.J. Russell, Optical frequency synthesizer for precision spectroscopy. Phys. Rev. Lett. **85**, 2264–2267 (2000)
11. T. Udem, R. Holzwarth, T.W. Hänsch, Optical frequency metrology. Nature **416**, 233–237 (2002)
12. M. Hentschel, R. Kienbergerlink, C. Spielmann, G.A. Reider, N. Milosevic, T. Brabec, P. Corkum, U. Heinzmann, M. Drescher, F. Krausz, Attosecond metrology. Nature **414**, 509–513 (2001)
13. E. Goulielmakis, M. Schultze, M. Hofstetter, V.S. Yakovlev, J. Gagnon, M. Uiberacker, A.L. Aquila, E.M. Gullikson, D.T. Attwood, R. Kienberger, F. Krausz, U. Kleineberg, Single-cycle nonlinear optics. Science **320**(5883), 1614–1617 (2008)
14. F. Krausz, M. Ivanov, Attosecond physics. Rev. Mod. Phys. **81**, 163–234 (2009)
15. C. Gohle, T.U.M. Herrmann, J. Rauschenberger, R. Holzwarth, H.A. Schuessler, F. Krausz, T.W. Hänsch, A frequency comb in the extreme ultraviolet. Nature **436**, 234–237 (2005)
16. A. Cingöz, D.C. Yost, T.K. Allison, A. Ruehl, M.E. Fermann, I. Hartl, J. Ye, Direct frequency comb spectroscopy in the extreme ultraviolet. Nature **482**, 68–71 (2012)
17. D. Bauer, I. Zawischa, D.H. Sutter, A. Killi, T. Dekorsy, Mode-locked Yb:YAG thin-disk oscillator with 41 μJ pulse energy at 145 W average infrared power and high power frequency conversion. Opt. Express **20**, 9698 (2012)
18. O. Pronin, J. Brons, C. Grasse, V. Pervak, G. Boehm, M.-C. Amann, V.L. Kalashnikov, A. Apolonski, F. Krausz, High-power 200 fs Kerr-lens mode-locked Yb:YAG thin-disk oscillator. Opt. Lett. **36**(24), 4746–4748 (2011)
19. T. Eidam, S. Hanf, E. Seise, T.V. Andersen, T. Gabler, C. Wirth, T. Schreiber, J. Limpert, A. Tünnermann, Femtosecond fiber CPA system emitting 830 W average output power. Opt. Lett. **35**(2), 94–96 (2010)
20. P. Russbueldt, T. Mans, J. Weitenberg, H.D. Hoffmann, R. Poprawe, Compact diode-pumped 1.1 kW Yb:YAG innoslab femtosecond amplifier. Opt. Lett. **35**(24), 4169–4171 (2010)
21. S. Naumov, A. Fernandez, R. Graf, P. Dombi, F. Krausz, A. Apolonski, Approaching the microjoule frontier with femtosecond laser oscillators, New J. Phys. 1, 216 (2005)
22. http://www.femtolasers.com/Oscillator.50.0.html
23. A. Stingl, M. Lenzner, C. Spielmann, F. Krausz, R. Szipöcs, Sub-10-fs mirror-dispersion-controlled Ti:sapphire laser. Opt. Lett. **20**(6), 602–604 (1995)
24. D.H. Sutter, G. Steinmeyer, L. Gallmann, N. Matuschek, F. Morier-Genoud, U. Keller, V. Scheuer, G. Angelow, T. Tschudi, Semiconductor saturable-absorber mirror assisted Kerr-lens mode-locked Ti:sapphire laser producing pulses in the two-cycle regime. Opt. Lett. **24**(9), 631–633 (1999)

25. A. Greborio, A. Guandalini, J.A. der Au, Sub-100 fs pulses with 12.5-W from Yb:CALGO based oscillators, SPIE. **8235**, 823511 (2012)
26. D. Bauer, F. Schättiger, J. Kleinbauer, D.H. Sutter, A. Killi, T. Dekorsy, Energies above 30 μJ and average power beyond 100 W directly from a mode-locked thin-disk oscillator, in *Advanced Solid-State Photonics*, Optical Society of America, Washington, DC, p. ATuC2 (2011)
27. C.R.E. Baer, C. Kränkel, C.J. Saraceno, O.H. Heckl, M. Golling, R. Peters, K. Petermann, T. Südmeyer, G. Huber, U. Keller, Femtosecond thin-disk laser with 141 W of average power. Opt. Lett. **35**(13), 2302–2304 (2010)

Chapter 2
Basics of Mode-Locking

This chapter is devoted to the theoretical basics underlying the mode-locking of an oscillator. Section 2.1 intuitively describes the concept of locking the oscillator modes and the different regimes of ultrashort pulse formation. Section 2.2 shows the role of different effects on pulse and generation. Specifically, the basics of soliton mode-locking and some important equations are formulated. Section 2.3 introduces the basic idea of pulse formation in the positive dispersion regime and advantages. Mode-locking instabilities are analyzed in Sect. 2.4 and a short summary finalizes this chapter.

2.1 Locking the Phases

Generally, the output power of an oscillator fluctuates as a consequence of the interference of different longitudinal modes with random phase relations as depicted in Fig. 2.1a. When the phases of the longitudinal modes become fixed, a single pulse can be generated in the resonator depicted in Fig. 2.1b. The randomly fluctuating modes of a He-Ne laser could be forced to oscillate with a fixed phase relationship via synchronous intracavity modulation by an acousto-optical modulator (AOM) as a first demonstration of mode-locking [1]. In the time domain, such a standing-wave AOM works as a shutter forming a time-window for the circulating pulse. In the frequency domain, this corresponds to the appearance of sidebands around the central optical frequency which have a fixed phase relation to all other sidebands; in other words, they are locked. This technique is an example of active mode-locking. Later it was found that a saturable absorber placed inside a cavity can also serve as a shutter [2], discriminating against CW and favouring pulsed operation. In that case there is no active modulator and intensity fluctuation modulates itself via a saturable absorber. In contrast to the previous technique this case can be classified as passive mode-locking. This manner of locking the cavity modes relies solely on the

O. Pronin, *Towards a Compact Thin-Disk-Based Femtosecond XUV Source*,
Springer Theses, DOI: 10.1007/978-3-319-01511-8_2,

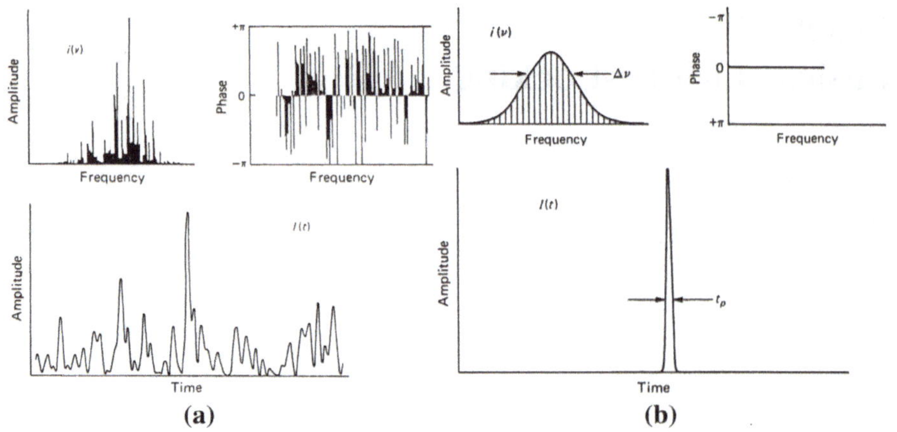

Fig. 2.1 **a** Output of a CW laser. The phases are randomly distributed, which results in a noisy intensity output. **b** Output of a mode-locked laser. The phases of different modes are locked, which results in a Fourier-limited pulse. The pictures are from [3]

intracavity-intensity dynamics and is passive in nature and is thus termed passive as opposed to active mode-locking.

Depending on the gain dynamics and the response time of the active modulator or saturable absorber, several scenarios of ultrashort pulse formation are possible. Figure 2.2a shows the situation when a time-window is formed for pulse build-up both by dynamic gain saturation (green) and a slow saturable absorber. This scenario can be realized experimentally for dye lasers with another dye as saturable absorber [4]. Another situation in Fig. 2.2b corresponds to the combination of a gain medium, the dynamics of which can be neglected on a short time scale, with a fast saturable absorber responsible for the formation of the time-window with net gain. A solid-state gain medium mode-locked with a dye saturable absorber or via KLM can be described in this way [2] for example. Mode-locking can also be realized without a

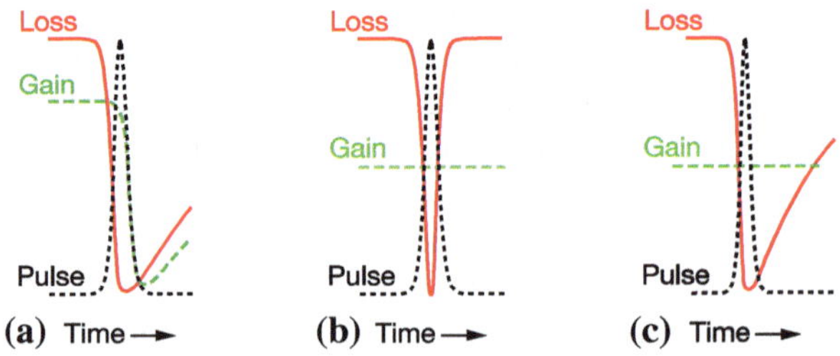

Fig. 2.2 Role of gain dynamics, response time of the active modulator and saturable absorber. **a** Mode-locking with a slow saturable absorber and dynamic gain saturation. **b** Mode-locking with a fast saturable absorber and without dynamic gain saturation. **c** Mode-locking with a slow saturable absorber and without dynamic gain saturation. The picture is from [6]

fast shutter and only with the help of intrinsically stable soliton pulses (Fig. 2.2c). This case describes mostly SESAM mode-locked solid-state gain-media [5]. In all cases from Fig. 2.2, when the contributions from nonlinearity and dispersion become crucial for pulse formation light solitons can be formed. Depending on the the sign of the intracavity dispersion, they can be categorized into the regimes of positive and negative dispersion and can both be described by the same complex nonlinear Ginzburg-Landau Eq. 2.12. Thus, independently of the mode-locking technique, understanding soliton pulse dynamics is important to gain insight into the different destabilizing mechanisms and prospects for energy scaling and achieving ultrashort pulse durations.

2.2 Basic Theory of Ultrashort Pulses

In the following, the effects influencing the generation and propagation of ultrashort pulses are summarized. An ultrashort pulse travelling inside the resonator experiences the combined action of gain, losses, self-phase modulation (SPM), self-amplitude modulation (SAM), dispersion and spectral filtering. This is displayed schematically in Fig. 2.3. $A(z, \tau)$ is the field envelope of the pulse, where z is the propagation distance and t is the local time (around the centre of the pulse).

2.2.1 *Gain and Loss*

The action of the gain and losses can be described as [7]

$$\frac{\partial A}{\partial z} = \frac{g_0 A(z, \tau)}{1 + E/E_s} - \ell A(z, \tau) = -\sigma A(z, \tau), \tag{2.1}$$

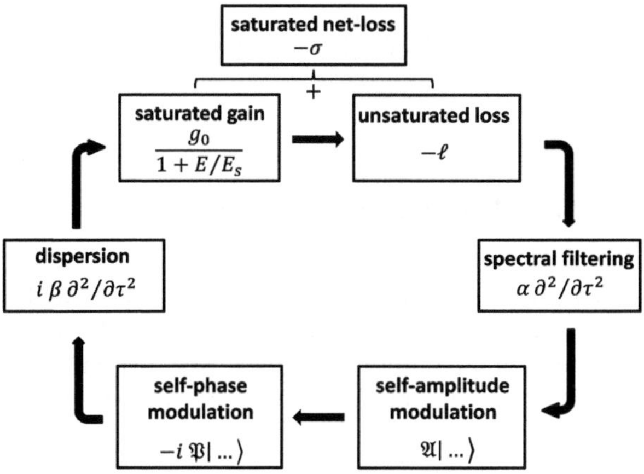

Fig. 2.3 Schematic of ultrashort pulse formation [7]

where g_0 is the unsaturated gain (small signal gain), E is the pulse energy and E_s is the saturation energy defined as $E_s = \hbar\omega_0 S/\sigma_g$ (σ_g is the gain cross-section, S is the beam area), ℓ is the linear power-independent loss. In Eq. 2.1 the gain does not depend on the time τ, thus steady-state gain saturation is considered here. This is a good approximation for the case of a solid-state oscillator when the typical intracavity energy is much smaller than the saturation energy, and especially for a TD oscillator when large spot sizes are used on the disk. The situation when gain and linear losses are equal corresponds to $\sigma = 0$, indicating marginal stability of the soliton. A value $\sigma > 0$ implies that only the soliton experiences the amplification and serves as a stability condition for soliton mode-locking.

2.2.2 Spectral Filtering

The spectrally dependent gain or loss works as a spectral filter and selects those frequencies which lie within the filter bandwidth with the profile $\Phi(\omega)$. In the frequency domain, it influences the pulse as follows:

$$\frac{\partial A(z,\omega)}{\partial z} = \Phi(\omega)A(z,\omega). \tag{2.2}$$

Assuming a Lorenzian shape of the $\Phi(\omega)$ and the filter bandwidth Ω being much larger than the corresponding bandwidth of the pulse $1/T$

$$\Omega \gg 1/T,$$

one can expand $\Phi(\omega)$ in a Taylor series in the frequency domain and transfer it to the time domain as [7]

$$\frac{\partial A(z,\omega)}{\partial z} \approx g(\omega_0)\left[1 - \frac{1}{\Omega}\frac{\partial}{\partial\tau} + \frac{1}{\Omega^2}\frac{\partial^2}{\partial\tau^2} - higher \quad order \quad terms\right]A(z,\tau). \tag{2.3}$$

2.2.3 Action of Self-Amplitude Modulation

Due to the mechanism of self-amplitude modulation the mode-locked regime is favoured over CW operation. In Chaps. 4 and 5 two main mechanisms for introducing self-amplitude modulation are described, namely SESAM and KLM. Both can be modelled as some effective saturable absorber with a response function

$$\frac{\partial A}{\partial z} = \mathfrak{F}(|A(z,\tau)|^2)A(z,\tau). \tag{2.4}$$

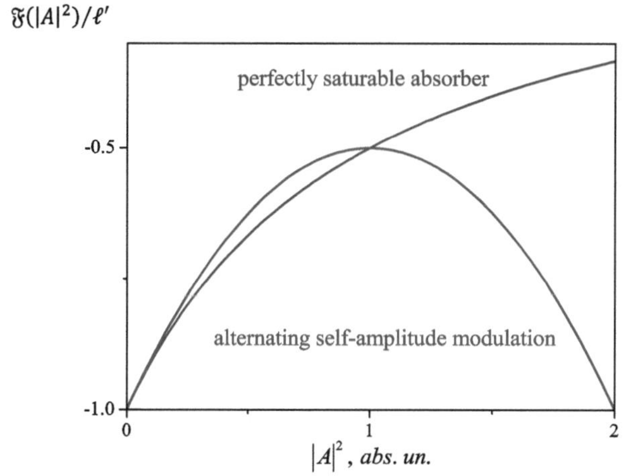

Fig. 2.4 Response functions of self-amplitude modulation [7]

An idealized perfectly saturable absorber has the response function shown in Fig. 2.4 (blue line), which depicts complete saturation of the absorber at increased power levels. The low power loss ℓ' approaches zero in this case. Mathematically, it can be described as [7]

$$\mathfrak{F}(|A(z,\tau)|^2) = -\frac{\ell'}{1+\kappa|A(z,\tau)|^2}. \tag{2.5}$$

This situation corresponds to hard-aperture KLM when a reduced beam size on the hard-aperture reduces the losses and also to an idealized SESAM (without TPA). The red curve in Fig. 2.4 shows that a decrease of the saturable losses ℓ' changes into an increase of the saturable losses at a certain value of the power. This can happen due to the nonlinearities in the SESAM, in particular TPA and in the case of soft-aperture KLM when an increase of the peak power tends to worsen the overlap of the beam with the pump spot. The analytic model is [7]

$$\mathfrak{F}(|A(z,\tau)|^2) = -\ell' + \kappa(1-\zeta|A(z,\tau)|^2)|A(z,\tau)|^2, \tag{2.6}$$

where κ is the inverse saturation power and ζ the coefficient of the self amplitude alteration. Both self-amplitude modulation behaviours can be considered valid when the response time of the saturable absorber is faster than the pulse duration. This is always true of the KLM mechanism (see Sect. 5.1), but may, however, not be true of a SESAM and sub-ps pulse durations (see Sect. 4.1). The oscillator described in Sect. 5.2 was mode-locked via hard-aperture KLM together with the contribution from a soft-aperture. Thus ideally both behaviours of self-amplitude modulation described by Eqs. 2.5 and 2.6 should in that case be included in the theoretical treatment.

2.2.4 Self-Phase Modulation

The refractive index n of a material depends on the incident electric-field intensity $I = 2n_0\epsilon_0 c|A(z,\tau)|^2$ [8]:

$$n = n_0 + n_2 I(\tau), \tag{2.7}$$

where n_2 is the nonlinear refractive index. During propagation in a medium of length L a pulse acquires the nonlinear phase

$$\phi_{NL} = \frac{2\pi}{\lambda} n_2 I(\tau) L = \frac{4\pi}{\lambda} n_2 L \frac{P}{S} = \gamma \frac{P}{S}, \tag{2.8}$$

where S is the beam area. The time-dependent phase corresponds to a time-dependent instantaneous frequency

$$\omega(\tau) = \omega_0 + \delta\omega(\tau),$$

where

$$\delta\omega(\tau) = \frac{\partial \phi_{NL}}{\partial \tau}.$$

The change in the instantaneous frequency is illustrated in Fig. 2.5 for sech^2 pulses propagating in a medium with positive n_2. The leading edge of the pulse is shifted to lower and the trailing edge to higher frequencies, thus obtaining a positive chirp.

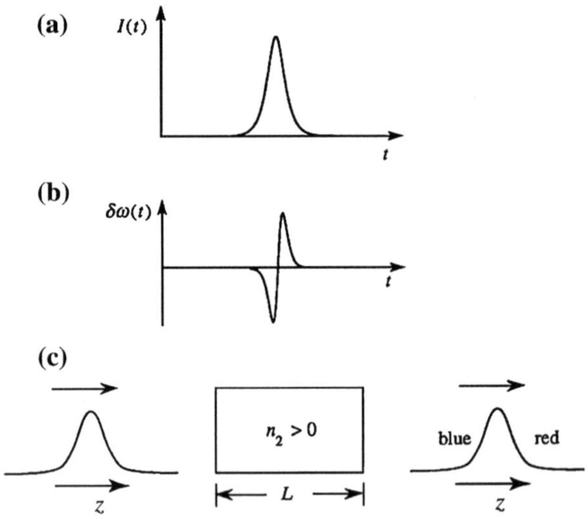

Fig. 2.5 Self-phase modulation. **a** Incident pulse. **b** Change in the instantaneous frequency of the pulse. **c** The leading edge of the pulse is shifted to lower and the trailing edge-to higher frequencies [9]

For high nonlinearities the frequency shift $\delta\omega$ may exceed the pulse bandwidth and broaden the spectrum. This effect is called self-phase modulation and will be demonstrated experimentally in Sect. 6.1. The contribution of self-phase modulation can be modelled according to [7]

$$\frac{\partial A}{\partial z} = -i\left[\gamma + \chi|A(z,\tau)|^2\right]|A(z,\tau)|^2 A(z,\tau). \tag{2.9}$$

The term $\chi|A(z,\tau)|^4$ serves as a correction for powers of higher order in $|A|$. It describes the influence of self-focusing on the self-phase modulation, in other words the dependence of γ on S.

2.2.5 Dispersion

Dispersion plays an important role in the generation of ultrashort pulses. The wave propagation constant $k(\omega)$ of a pulse is frequency-dependent and can be expanded in a Taylor series as

$$k(\omega) = k(\omega_0) + k_1(\omega - \omega_0) + \frac{1}{2}k_2(\omega - \omega_0)^2 + \textit{higher order terms}, \tag{2.10}$$

where $k(\omega_0)$ represents a phase shift, $k_1 = \partial k/\partial\omega = 1/v_g$ is the inverse group velocity and $k_2 = (\partial/\partial\omega)(1/v_g)$ is the group velocity dispersion. The latter multiplied by the cavity length is the group delay dispersion (GDD) coefficient $\beta \equiv L_{cav}k_2$. In dispersive media, different parts of the optical spectrum propagate with different group velocities, effectively spreading the pulse. A medium with normal dispersion delays the high-frequency part of the spectrum with respect to the low-frequency part and causes a positive chirp. By definition, the GDD coefficient β is positive in the case of normal dispersion and negative for anomalous dispersion. Obviously, the broader the spectrum of an ultrashort pulse the stronger the influence of higher-order terms. Dispersion compensation can be accomplished by an arrangement consisting of prisms [10], gratings [11] or dispersive mirrors [12]. Discovery of the latter had a drastic impact on ultrashort pulse generation due to the ability to control higher-order dispersions and due to its compactness. For more details on the dispersive mirrors used in this work see [13] and Sect. 4.2. The coefficients k and k_1 can be included in the definition of the wave number and local time τ, respectively. Thus, the influence of the GDD is described as

$$\frac{\partial A}{\partial z} = i\beta\frac{\partial^2 A(z,\tau)}{\partial\tau^2} + \textit{higher order terms}. \tag{2.11}$$

2.2.6 Haus Master Equation

Due to the small changes experienced by the pulse during one round trip in the cavity the different contributions formulated in Eqs. 2.1, 2.3, 2.4, 2.9, 2.10 can be joined together:

$$\frac{\partial A(z,\tau)}{\partial z} = \left[-\sigma(E) + \frac{g(\omega_0)}{\Omega^2}\frac{\partial^2}{\partial \tau^2} + \mathfrak{F}(|A|^2)\right] A(z,\tau) + i\left[\beta\frac{\partial^2}{\partial \tau^2} - \gamma|A|^2 - \chi|A|^4\right] A(z,\tau). \tag{2.12}$$

Equation 2.12 is the modified Haus master equation from [14] and can be treated as a complex nonlinear Ginzburg-Landau equation for which exact solutions are known only in a few cases [7]. Equation 2.12 can be considered an approximation of the small self-amplitude modulation, by neglecting dissipative effects, gain and spectral filtering:

$$\frac{\partial A(z,\tau)}{\partial z} = i\beta\frac{\partial^2}{\partial \tau^2} A(z,\tau) - i\gamma|A|^2 A(z,\tau) \tag{2.13}$$

Equation 2.13 is a nonlinear Schrödinger (NLS) equation [15] which, in the case of negative GDD, has an exact solution, a soliton pulse,

$$A(z,\tau) = \sqrt{P_0} sech(\tau/T), \tag{2.14}$$

where P_0 is the peak power for a pulse of width T. The FWHM pulse duration $\Delta\tau$ is related to T as $\Delta\tau = 2ln(1+sqrt(2))T \approx 1.76T$ [15]. Between the soliton energy E, pulse width T and GDD value β there are the simple relations

$$T = \sqrt{|\beta|/\gamma P_0}, \qquad E = 2|\beta|/\gamma T. \tag{2.15}$$

These relations describe well the propagation of optical pulses in anomalous dispersion fibres [15]. However, one has to be careful in applying them to the oscillator because of the many strict approximations made to formulate the NLS Eq. 2.13 and the inability to define a soliton stability range in terms of E and GDD. Due to instabilities the peak power P_0 cannot be higher than a certain threshold value P_{th} related to the pulse energy as

$$E = 2\sqrt{P_{th}|\beta|/\gamma}. \tag{2.16}$$

Thus an increase of the GDD stretches the pulses, which leads to reduced peak power at higher pulse energies. In other words, one has to increase the GDD in order to increase the pulse energy but has to pay for it with an increased pulse duration. In Chaps. 4 and 5 experiments on SESAM and Kerr-lens mode-locking in the regime of negative dispersion are described and the aforementioned effects are illustrated.

2.2.7 *Limits of the Pulse Duration*

The question of the minimal achievable pulse duration from the oscillator within a certain parameter range is difficult to answer. It strongly depends on the method of mode-locking and the characteristics. To answer this numerical simulations of Eq. 2.12 are needed. It is interesting that Eq. 2.12 does not set the gain bandwidth (spectral filtering) as an ultimate upper limit for the minimal achievable pulse duration (in the regime of negative dispersion). Thus, with a rather high SAM coefficient ℓ' it is possible to exploit spectral broadening due to SPM in order to reduce the pulse duration. This was done in, for example, [16, 17] with bulk KLM Yb:Lu_2O_3 and Ti:Sa oscillators, respectively. In the case of high-power/energy operation, however, the emission-bandwidth-limited oscillator performance is a reasonable and good goal to aim for.

Spatial hole burning can also be a favourable factor for achieving minimum pulse duration [18, 19] as a consequence of the effective flattening of the gain spectrum. Dispersion management and Kelly sidebands (formation of dispersive waves) [20, 21] become very important as soon as the generation of ultrashort pulses with only a few optical cycles is aimed for.

2.3 Soliton Mode-Locking with Positive Dispersion

In the negative dispersion regime pulses are chirp-free. This is a natural consequence of soliton mode-locking when SPM introduces a positive chirp and the negative GDD compensates for this chirp. The balance between nonlinearity and dispersion forms the soliton and keeps it stable. The complex nonlinear Ginzburg-Landau Eq. 2.12 can, however, be solved for positive GDD ($\beta > 0$), leading to the chirped solution [14, 7]

$$A(z, \tau) = \sqrt{P_0} sech(\tau/T)^{(1+i\psi)}, \tag{2.17}$$

where ψ is the chirp, which means that the pulse is stretched ψ times in comparison with the Schrödinger soliton 2.14. This has several important consequences:

- the peak power inside the oscillator is reduced by the factor ψ. ψ can be as high as 100
- as a result the GDD is approximately an order of magnitude smaller than that needed in the regime of negative GDD [7]
- increase of the energy leads to a stronger chirp ψ

The physical explanation for the existence of the positive dispersion regime is as follows: SPM (Eq. 2.9) causes substantial spectral broadening and chirping. On the other hand, spectral filtering (limited gain bandwidth, Eq. 2.3) cuts off this spectrum at the edges and consequently shortens the pulses. The interplay between those two factors supports a self-sustained "chirped soliton". For more details on the theoretical description of this regime applied to oscillators see [7] and for experimental realization see Sect. 5.4.

2.4 Mode-Locking Instabilities

There are several types of mode-locking instabilities: the onset of multi-pulsing, Q-switched mode-locking (QML), simultaneous lasing in CW and combinations of these. The example of QML is shown in Fig. 2.6. The pulse train is modulated at a somewhat low frequency which is defined by the fluorescence lifetime of the gain medium and other oscillator parameters [22]. Not necessarily the pulse train should be strongly periodically modulated. Generally, the envelope of the pulse train can be of a random shape with a randomly large modulation amplitude. The origin of this instability lies in the fact that an increase of the peak power causes stronger absorber bleaching up to the point when the gain and absorber become well saturated. QML was studied in detail in [23, 22] for SESAMs as slow saturable absorbers:

$$E^2 > E_{gain} E_{abs} \ell', \tag{2.18}$$

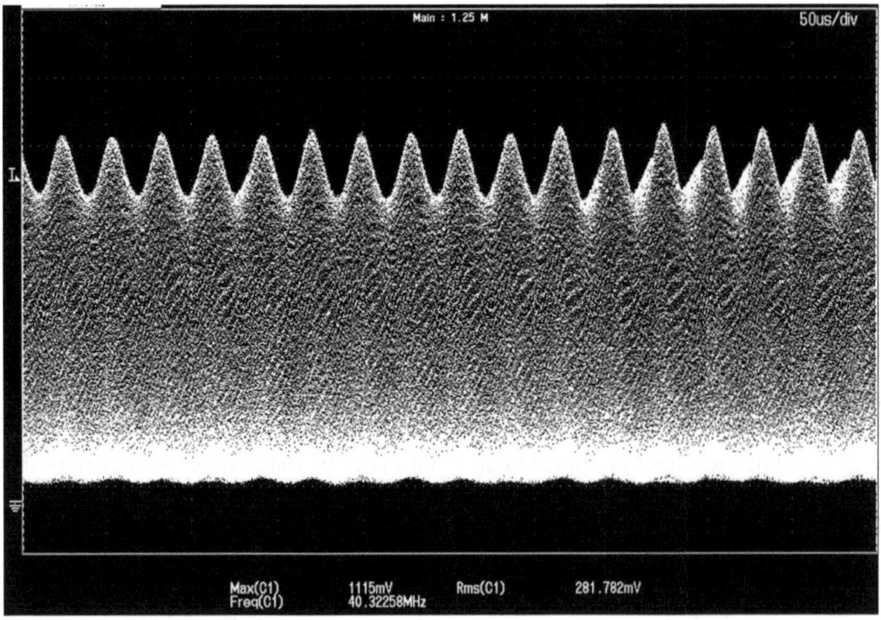

Fig. 2.6 Measured in experiment periodically modulated pulse train from SESAM-ML oscillator in Sect. 4.4. Scale division is 50 μs

where E_{gain} and E_{abs} are the saturation energy of the gain medium and absorber, respectively. The formula implies that the pulse energy squared should be larger than the product of the saturation energies and the absorber modulation depth ℓ' in order to suppress QML. It is worth noting that QML was considered in [23] as a "deviation" regime from the stable steady-state ML. The onset of QML during oscillator start-up and its dynamics is beyond the scope of this description. However, as was found experimentally (see Sect. 4.4) it is this onset of QML which prevents moving to the regime of stable ML with a SESAM. For clarity, Q-switched mode-locking will be separated into self-Q-switching, chaotic QML and QML just in the frame of this work. The latter corresponds to the "classical case" of strongly periodic modulation depicted in Fig. 2.6. The description of self-Q-switching and chaotic QML is given in experimental Sect. 4.4.

The QML instability for the case of KLM was also studied experimentally [24] and theoretically [25], but mostly for Ti:Sa bulk oscillators with the rather short relaxation time of $<5\,\mu$s. The saturation behaviour of soft-aperture KLM in Fig. 2.4 is most likely the main reason for the suppression of QML in the experiments.

When the intracavity peak power increases, a single circulating pulse can break up into two (double pulsing) and more pulses. This can happen when TPA, for instance, becomes pronounced in SESAM or due to the excessive SPM which cannot be balanced by the introduced GDD. This regime is illustrated in Fig. 5.5. The soliton can also transfer its excess energy into a dispersive wave or the continuum (CW). The latter scenario is shown in Fig. 4.14.

2.5 Summary

The basics of soliton mode-locking are described. Factors such as SPM, dispersion, spectral filtering and different mechanisms of self-amplitude modulation are responsible for the formation of solitons and govern these according to the complex nonlinear Ginzburg-Landau equation. This briefly introduced theoretical description together with the experimental Chaps. 4 and 5 gives a more unified picture of soliton mode-locking and helps to understand the main assumptions in the model.

References

1. L.E. Hargrove, R.L. Fork, M.A. Pollack, Locking of He-Ne laser modes induced by synchronous intracavity modulation. Appl. Phys. Lett. **5**(1), 4–5 (1964)
2. A. DeMaria, J. Glenn, W.H., M. Brienza, and M. Mack, "Picosecond laser pulses". Proc. IEEE **57**(1), 2–25 (1969)
3. W. Koechner, *Solid-State Laser Engineering*, 6th edn. (Springer, New York, 2006)
4. E. Ippen, C. Shank, A. Dienes, Passive mode locking of the cw dye laser. Appl. Phys. Lett. **21**(8), 348–350 (1972)

5. F. Kärtner, J. der Au, U. Keller, Mode-locking with slow and fast saturable absorbers-what's the difference? IEEE J. Sel. Top. Quantum Electron. **4**(2), 159–168 (1998)
6. U. Keller, Ultrafast solid-state laser oscillators: a success story for the last 20 years with no end in sight. Appl. Phys. B **100**, 15–28 (2010)
7. V.L. Kalashnikov, *Solid state lasers*, ch. Chirped-Pulse Oscillators: Route to the Energy-Scalable Femtosecond Pulses. (InTech, 2012), pp. 145–184
8. S.A. Akhmanov, SYu. Nikitin, *Physical Optics* (Clarendon Press, Oxford, 1997)
9. R.W. Boyd, *Nonlinear Optics*, 3rd edn. (Academic Press, Boston, 2007)
10. R.L. Fork, O.E. Martinez, J.P. Gordon, Negative dispersion using pairs of prisms. Opt. Lett. **9**(5), 150–152 (1984)
11. E. Treacy, Optical pulse compression with diffraction gratings. IEEE J. Quantum Electron **5**(9), 454–458 (1969)
12. R. Szipöcs, K. Ferencz, C. Spielmann, F. Krausz, Chirped multilayer coatings for broadband dispersion control in femtosecond lasers. Opt. Lett. **19**(3), 201–203 (1994)
13. V. Pervak, O. Pronin, O. Razskazovskaya, J. Brons, I.B. Angelov, M.K. Trubetskov, A.V. Tikhonravov, F. Krausz, High-dispersive mirrors for high power applications. Opt. Express **20**(4), 4503–4508 (2012)
14. H.A. Haus, J.G. Fujimoto, E.P. Ippen, Structures for additive pulse mode locking. J. Opt. Soc. Am. B **8**(10), 2068–2076 (1991)
15. G.P. Agrawal, *Nonlinear Fiber Optics*, 3rd edn. (Academic Press, New York, 2001)
16. M. Tokurakawa, A. Shirakawa, K. ichi Ueda, H. Yagi, S. Hosokawa, T. Yanagitani, A.A. Kaminskii, Diode-pumped 65 fs Kerr-lens mode-locked Yb^{3+}:Lu_2O_3 and nondoped Y_2O_3 combined ceramic laser. Opt. Lett. **33**(12), 1380–1382 (2008)
17. R. Ell, U. Morgner, F.X. Kärtner, J.G. Fujimoto, E.P. Ippen, V. Scheuer, G. Angelow, T. Tschudi, M.J. Lederer, A. Boiko, B. Luther-Davies, Generation of 5-fs pulses and octave-spanning spectra directly from a Ti:sapphire laser. Opt. Lett. **26**, 373–375 (Mar 2001)
18. B. Braun, K.J. Weingarten, F.X. Kärtner, U. Keller, Continuous-wave mode-locked solid-state lasers with enhanced spatial hole burning. Appl. Phys. B **61**, 429–437 (1995)
19. R. Paschotta, J. Aus der Au, G. Spühler, S. Erhard, A. Giesen, U. Keller, Passive mode locking of thin-disk lasers: effects of spatial hole burning. Appl. Phys. B **72**, 267–278 (2001)
20. S. Kelly, Characteristic sideband instability of periodically amplified average soliton. Electron. Lett. **28**(8), 806–807 (1992)
21. M. Dennis, I. Duling III, Experimental study of sideband generation in femtosecond fiber lasers. IEEE J. Quantum Electron. **30**(6), 1469–1477 (1994)
22. A. Schlatter, S.C. Zeller, R. Grange, R. Paschotta, U. Keller, Pulse-energy dynamics of passively mode-locked solid-state lasers above the Q-switching threshold. J. Opt. Soc. Am. B **21**(8), 1469–1478 (2004)
23. C. Hönninger, R. Paschotta, F. Morier-Genoud, M. Moser, U. Keller, Q-switching stability limits of continuous-wave passive mode locking. J. Opt. Soc. Am. B **16**(1), 46–56 (1999)
24. Q. Xing, W. Zhang, K. Yoo, Self-Q switched self-mode-locked Ti:sapphire laser. Opt. Commun. **119**(12), 113–116 (1995)
25. J. Jasapara, W. Rudolph, V.L. Kalashnikov, D.O. Krimer, I.G. Poloyko, M. Lenzner, Automodulations in Kerr-lens mode-locked solid-state lasers. J. Opt. Soc. Am. B **17**(2), 319–326 (2000)

Chapter 3
High-Power Thin-Disk Resonator and Gain Medium

Progress in the development of laser diodes has attracted a lot of interest to Yb-doped gain media over the last few decades. Such gain media are of a quasi-three-level nature, and have a narrow absorption bandwidth, previously considered as undesirable, but then turned out to be advantageous with the availability of powerful high-brightness pump diodes. Together with the invention of the thin-disk concept [1] and routine growth of Yb:YAG crystals [2] a new class of thin-disk solid-state diode-pumped lasers was established [3]. These systems are truly power-scalable, have large mode areas over the optical elements and are sensitive to misalignment.

Section 3.1 introduces the thin-disk geometry and its potential for power scaling. Promising gain materials and their properties are briefly described in Sect. 3.2. Section 3.3 considers stability, influence of a thermal lens and misalignment sensitivity of high-power resonators.

3.1 Thin-Disk Concept

To date, fibre and thin-disk lasers constitute the main part of high average power lasers reaching multi-kW powers in the multimode CW regime [4, 5]. This outstanding performance is a consequence of the good heat transport geometry. Fibres have a long and small-diameter gain medium with a large ratio of surface to active volume. High nonlinearities, on the other hand, prevent reaching high peak intensities due to the small light-guiding region in such systems and call for more bulky amplifier configurations. In contrast, the thin-disk geometry utilizes very thin ($<200\,\mu m$) and large-diameter (>5 mm) laser crystals mounted with solder or glue [6] to a water-cooled heat sink (Fig. 3.1a). The ratio of the disk diameter to the thickness is large, providing very efficient cooling and nearly one-dimensional, heat flow along the beam axis (see Fig. 3.1a). Consequently, this geometry results in a low transversal thermal gradient, which in turn is a main prerequisite for good beam quality. The back surface of the disk is HR-coated for both pump and laser wavelengths and the front

O. Pronin, *Towards a Compact Thin-Disk-Based Femtosecond XUV Source*,
Springer Theses, DOI: 10.1007/978-3-319-01511-8_3,

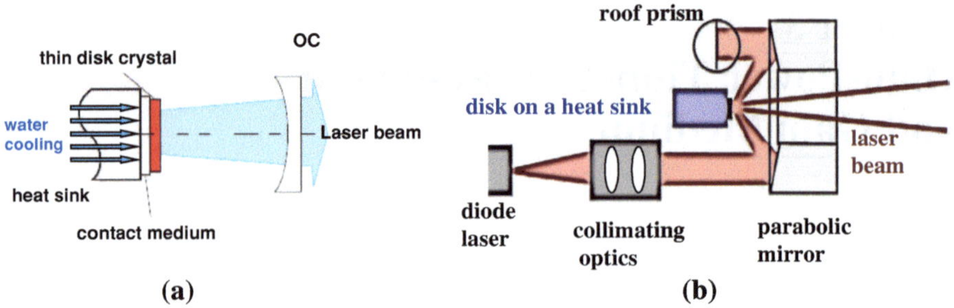

Fig. 3.1 Schematic of the thin-disk (**a**) and multi-pass disk pumping geometry (**b**) adopted from [7]

surface is AR-coated also for both wavelengths. Obviously, the pump absorption in the TD is low for reasonable crystal doping levels. Thus, a multi-pass configuration with the disk placed in the focus of a parabolic mirror (Fig. 3.1b) is standardly used. As demonstrated in Fig. 3.1b, the pump radiation is coupled into a fibre to provide excellent homogenization, propagates through the collimating glass optics and is then focused onto the disk by means of a parabolic mirror. The parabolic mirror together with a roof-prisms allow many passes to be realized. Depending on the geometry, over 40 passes through the gain medium can be achieved with the pump absorption well above 90 %.

Apparently, the thinner the disks the better is the heat dissipation. The pump absorption can be kept constant by choosing a material with higher doping level and reducing its thickness. Unfortunately, the doping level cannot be strongly increased because of the concentration quenching and reduced thermal conductivity [8]. Thus, these two issues should always be addressed in choosing a crystal for thin-disk operation.

3.1.1 Thin-Disk and Module in this Work

For all experiments described in this thesis a thin-disk laser head TDM 1.0 from *D+G GmbH* was used [9]. It is aligned for 24 passes of the pump beam through the gain medium. Two similar Yb:YAG thin-disks soldered to CuW heat sinks with 7 % at. doping level were verified experimentally (see Table 3.1)

3.1.2 Power Scaling

The importance of the TD geometry lies in its "true power scalability". The term power scalability can be defined according to [10, 11]

Table 3.1 The specifications of the two thin-disks used in this work (from *D+G GmbH*)

Thin-disk sample	TD 157	TD 159
Yb concentration, at. %	7	7
Free aperture, mm	9	9
Thickness, μm	215	215
R_x, m	5.6	5.9
R_y, m	6.4	6.5
Treatment	Edge bevelled	Edge bevelled

Well-defined systematic scaling procedure which makes it possible to increase substantially and repeatedly the output power without making the main problems more severe.

For instance, reduction of the quantum defect by pumping at the zero-phonon line and applying cryogenic cooling leads to an increase of the output power of a laser system. These measures do not, however, make the laser system "power-scalable". In contrast, the thin-disk geometry allows an increase of the output power by a factor N by applying N times higher pump power to a $\sqrt{N}$ times larger pump spot or, in other words, by enlarging the pump spot on the disk in proportion to the pump power. This scaling procedure has near-zero consequences on the thermal lens of the disk, thanks to the effective cooling geometry. Certainly, this approach is valid in a relatively large power range and at some point may become limited , e.g., due to amplified spontaneous emission or some practical issues, e.g., misalignment sensitivity of the power-upscaled resonator. An increase of the number of TDs in the resonator can also be considered a method of power scaling, but is less practicable (just being costly).

Understanding of the power scaling principle is important as it will allow a deeper insight into the limitations of mode-locked oscillators and afford promising prospects of their future development.

3.2 Yb:YAG and Yb:Lu_2O_3

The range of different solid-state gain media for femtosecond operation is very broad. It includes transition-metal-doped materials such as Ti:sapphire, Cr^{2+}-doped crystals, rare earth-doped Nd^{3+}, Yb^{3+}, Tm^{3+} crystals and many others [3]. The article [13] gives a detailed overview of different gain media and their suitability for high-power TD operation. According to this paper the general requirements for the "ideal" gain medium can be summarized as follows:

1. broad emission bandwidth $\Delta\lambda_{em}$ for femtosecond operation
2. absorption wavelength range $\Delta\lambda_{abs}$ suitable for pumping with laser diodes
3. small quantum defect $q = 1 - \lambda_{abs}/\lambda_{em}$ to minimize the heat load
4. absence of parasitic effects such as excited state absorption, up-conversion or cross-relaxation
5. high emission and absorption cross-sections

6. high thermal conductivity
7. high doping concentrations
8. weak thermo-optical coefficient dn/dT
9. good mechanical and thermo-mechanical properties

Other aspects such as the availability of crystals in large sizes, their optical quality and anisotropic behaviour of thermo-optical and thermo-mechanical properties are also very important. The laser gain media considered in this work are limited to those already demonstrated in femtosecond operation and with output powers above 10 W (see Fig. 1.2 and Table A.1). These are experiments with Yb-doped garnet, sesquioxide and tungstate materials: Yb:YAG; Yb:Lu_2O_3, Yb:$LuScO_3$; Yb:KLu$(WO_4)_2$, Yb:KY$(WO_4)_2$. The main disadvantage of the KY$(WO_4)_2$ and KLu$(WO_4)_2$ host crystals strong anisotropic thermo-optical and thermo-mechanical properties. Yb:$LuScO_3$ has a rather broad emission bandwidth of 22 nm [14] but unfortunately also a low thermal conductivity $\kappa_{LuScO_3} < 4$ W/mK [14] and a very difficult growth process which is related to the high melting temperature $T_m = 2450\,°C$. This difficult growth limits the crystal quality and prevents production in large sizes. The growth of Yb:Lu_2O_3 is as difficult as Yb:$LuScO_3$, but it has better thermal conductivity $\kappa_{Lu_2O_3} > 12$W/mK which changes only slightly with the doping level.

As can be seen, the constraints in obtaining high average power and femtosecond pulses in the thin-disk geometry leave out Yb^{3+}-doped crystals and two most promising materials Yb:YAG and Yb:Lu_2O_3. So far Yb:YAG is the most widespread, well-established solid-state laser material commercially available in excellent optical quality and large crystal sizes. It has a broad absorption line at 940 nm, as well as a zero-phonon line at 969 nm (Fig. 3.2a) suitable for pumping by wavelength-stabilized high-power diodes [15]. As a consequence of its low quantum defect, absence of excited-state absorption and good thermal conductivity Yb:YAG is mainly [16] chosen over all other materials in thin-disk lasers. The only drawback of Yb:YAG is its relatively low emission bandwidth around $\Delta\lambda_{em,Yb:YAG} \approx 9$ nm (Fig. 3.2b). Indeed, the bandwidth is only slightly narrower than that for Yb:Lu_2O_3 with $\Delta\lambda_{em,Yb:Lu_2O_3} \approx 13$ nm. Both Yb:YAG and Yb:Lu_2O_3 were used for the mode-locking experiments in this work (see Sects. 5.2 and 5.5) and have similar properties, as can be seen in Table 3.2. The best results were obtained with Yb:YAG. Questions of the optical quality of the Yb:Lu_2O_3 crystals available proved to be one of the limiting factors in the mode-locking experiments (see Sect. 5.5). A short summary of the most important properties of these two materials is presented in Table 3.2.

3.3 Resonator Design

The resonator design for high-power operation is challenging. Even though the TD concept [1] eliminates the thermal lens of the gain media to some extent and TD CW lasers in fundamental TEM_{00} operation with output powers exceeding 500 W [21]

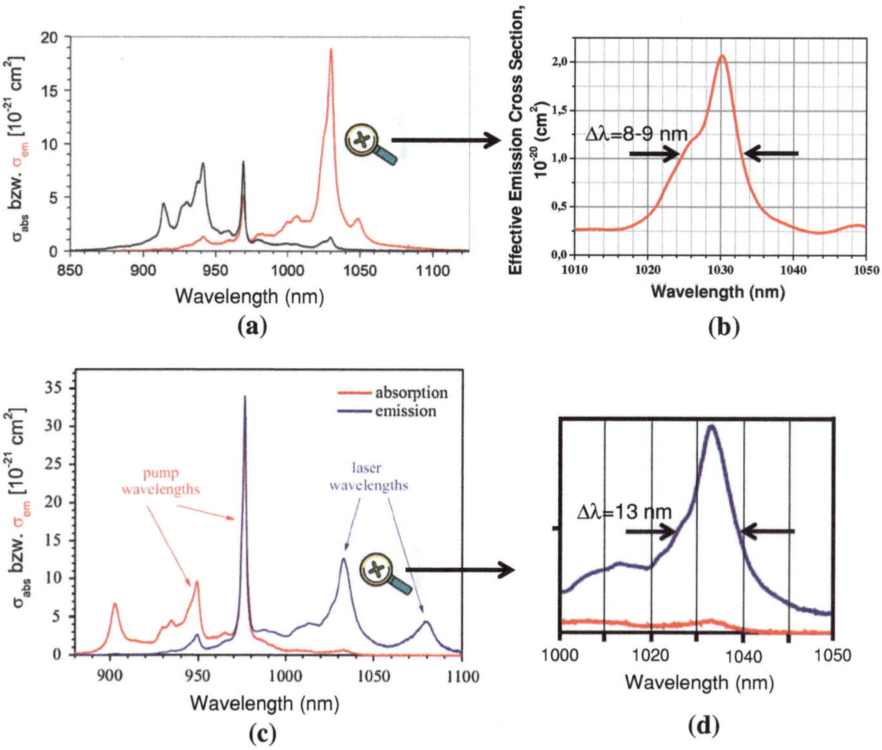

Fig. 3.2 Emission and absorption spectrum of Yb:YAG (**a**, **b**) [6, 12] and Yb:Lu_2O_3 (**c**, **d**) [2]

Table 3.2 Different characteristics of Yb:YAG and Yb:Lu_2O_3 gain media

Characteristic	Yb:YAG	Yb:Lu_2O_3
Emission wavelength, nm	1,030 [6]	1,033 [2]
Emission bandwidth (FWHM), nm	8–9 [6, 17]	13 [2, 17]
Emission crossection, 10^{-21} cm^2	20 [12]	12 [2]
Absorption wavelength, nm	940; 969 [12]	976 [2]
Absorption bandwidth (FWHM), nm	18; 2 [8]	2 [18]
Absorption crossection, 10^{-21} cm^2	8 [8]	30 [2]
Fluorescence lifetime, ms	0.95 [19]	0.82 [20]
Thermal conductivity un-doped, W/mK	9.5 [8]	12.6 [2]
Thermal conductivity Yb doped with 10 at. %, W/mK	6 [8]	11 [2]

were demonstrated, mode-locked operation was limited to the appreciably lower output powers of 150 W [18, 22]. The reason for this is the several additional elements necessary for soliton mode-locking: mode-locker and chirped mirrors. The heat dissipated in SESAM (as a mode-locker) and chirped mirrors causes pronounced thermal effects limiting the performance of the mode-locked oscillator (see measurements in Sect. 4.1). The cavity should be designed to tolerate these thermal lenses. In fact, there are more constraints on the cavity design in mode-locked operation arising from high intracavity peak intensities. Such peak intensities have to be avoided by designing a cavity with large mode areas on all intracavity elements susceptible to damage. As will be shown further on, the two requirements to tolerate several intracavity thermal lenses and have large mode areas along the intracavity optics are in slight contradiction. Moreover, they can be fulfilled with just a certain trade-off on the misalignment sensitivity of the designed resonator. The design of resonators for mode-locked high power oscillators is subject to the following requirements:

- CW fundamental TEM_{00} operation at maximum pump (output) power with all intracavity components necessary for mode-locked operation such as SESAM, dispersive mirrors, Kerr medium, Brewster plate, AOM etc ;
- mode sizes on intracavity elements should be chosen such as to keep the intensity below the damage threshold;
- small misalignment sensitivity

This section describes these requirements and trade-offs in more detail.

3.3.1 Cavity Stability

Every element inside the cavity can be represented by its own ABCD matrix and the whole cavity by multiplication of these matrix elements. It is convenient to introduce a complex Gaussian beam parameter q, which describes the Gaussian beam as [23, 24]

$$\frac{1}{q} = \frac{1}{R} + \frac{i\lambda}{\pi\omega^2}. \tag{3.1}$$

Thus, the real part in Eq. 3.1 determines the radius of curvature R of the phase front of the Gaussian beam and the imaginary part determines the beam radius ω (as 1/e^2 of intensity drop) for a certain wavelength λ. Together with the known cavity transfer matrix $M = \begin{pmatrix} A & B \\ C & D \end{pmatrix}$ the Gaussian beam changes as

$$q_{i+1} = \frac{q_i A + B}{q_i C + D}. \tag{3.2}$$

By definition, the cavity is stable when a Gaussian beam reproduces itself after one round trip $q_i = q_{i+1} = q$. In this case Eq. 3.2 can be simplified to a quadratic equation $Cq^2 + (D - A)q - B = 0$.[1] The radia ω and R can be calculated after comparison with Eq. 3.1. The discriminant $(D - A)^2 + 4BC$ must be negative to keep q a complex number; this together with $det[M] = 1$ implies

[1] The solutions of the equation provide q parameter expressed through the matrix elements.

$$\frac{(A+D)^2}{4} < 1. \tag{3.3}$$

The relation serves as a definition for resonator stability. The matrix elements A, D should be calculated to evaluate resonator stability. Values 0 and 1 in Eq. 3.3 correspond to marginal cavity stability. However, it may be unclear to the experimentalist at which stability values the oscillator should operate: 1, 0.5 or even 0.01? This and the following questions: "Can the stability edge of the cavity be reached in the experiment or will the oscillator still lase near the edge?", are reasonable. The answer lies in the treatment of the cavity stability in close conjunction with the cavity misalignment. A less stable cavity means it is most susceptible to misalignment and as long as constraints on the cavity alignment stay realistic and can be satisfied by the overall mechanical stability of the oscillator (precise and stable mirror mounts, vibration elimination and damping) the stability edge can be reached. However, this situation is rather idealistic and does not include the effect of the gain in the laser and is more valid for passive enhancement cavities. A more realistic scenario is that the beam profile strongly changes at the stability edge and causes poor overlap with the pump spot. Because of this, further approaching the stability edge interrupts laser operation. Often weak cavity astigmatism becomes strongly pronounced near the stability edge and the cavity becomes unstable at either the tangential or sagittal[2] plane. These issues are readily observable in the KLM oscillator, described in Sect. 5.2.

An interesting issue is mode degeneracy at the stability edge, which allows one to mode-lock higher transverse modes of the oscillator. This is discussed in Sect. 5.3.4 in the light of mode quality degradation of the KLM thin-disk oscillator.

Originally, the cavity design specified here was calculated by means of the Winlase commercial software, based on ABCD-matrix formalism. Later on, a more flexible and advanced code for cavity calculations and misalignment characterization was developed by Jonathan Brons. Figures 3.3b and 3.5b were calculated with this code.

3.3.2 Influence of a Thermal Lens

Since the first developments on high-power solid-state CW lasers, thermal lens stands out as the most crucial limitation on output power. This problem was partially solved by inventing and realizing a TD concept. In fact, there are many other elements inside the cavity which are not of TD geometry and therefore susceptible to thermal lensing. A thermal lens is caused by the dependence of the material's refractive index on temperature, by thermal expansion of the material as well as by mechanical deformations of the whole structure (e.g., heat sink) which supports the material [25]. The thermally induced temperature profile can be of complex shape, but to the first order it is of parabolic profile, thus causing a spherical thermal lens (see analogy to Kerr lens in Sect. 5.1). The compensation of the spherical part can be performed by

[2] The tangential plane is parallel to an optical table and the sagittal plane is perpendicular.

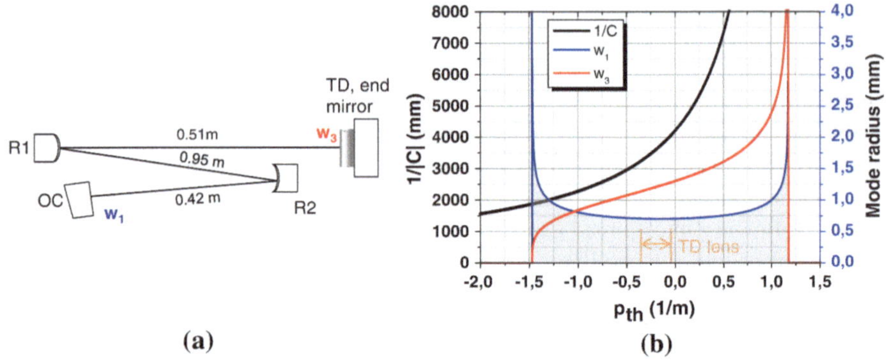

Fig. 3.3 **a** Simple resonator for CW TEM_{00} operation. The TD is placed as a cavity end mirror to minimize wavefront distortions. $R_{TD} \sim 6\,m$, $R1 = -4\,m$; $R2 = 4\,m$; **b** Misalignment sensitivity (1/C), mode radius on the TD (w_3), mode radius on the OC (w_1) vs dioptric power of the TD. The region marked by orange arrow corresponds to the expected TD thermal lens

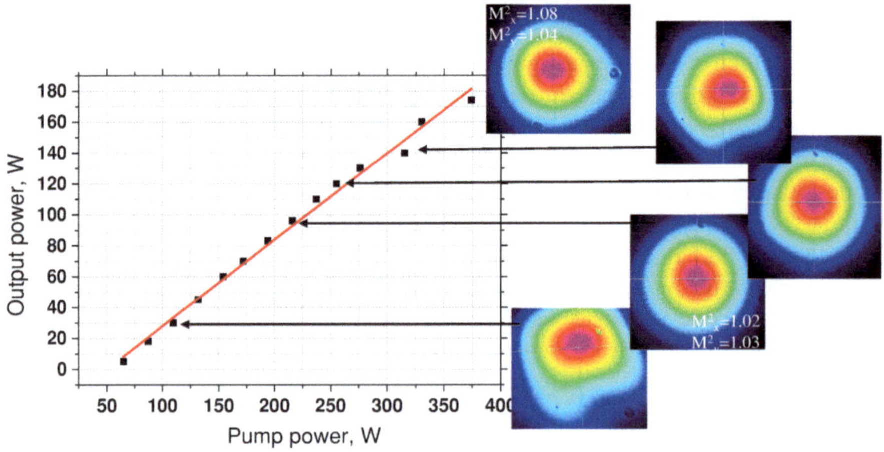

Fig. 3.4 Output power vs pump power for the resonator in Fig. 3.3. Experimental beam profile pictures are taken for different pump power levels

inserting a second lens[3] of opposite sign [25], using a deformable mirror [26, 27] or by implementing a telescope inside the cavity [28].

The influence of the thermal lens on the resonator stability was addressed by Magni [29]. To any complex cavity one can find the equivalent resonator. A thermal lens variation leads to destabilization of the cavity and consequently several stability zones appear in this analysis. In order to illustrate the above-mentioned, a CW cavity with only one element exhibiting a thermal lens, the TD, is considered. As specified by the

[3] in high power oscillator the use of transmissive optics is undesirable, therefore mirrors play role of lenses

supplier (Table 3.1) the initial (unpumped) radius of curvature of the TD is ~6 m. The change of focal power per 100 W pump power is 0.11/m. This value is approximate as it depends on the absolute pump power, the size of the pump spot and the output coupling rate. The disk head was supplied with a collimating optics to provide a pump spot diameter of 3.2 mm. The recommended pump power density for the TD is 4 kW/cm^2, which corresponds to 400 W maximum pump power. The fundamental mode diameter is chosen as 2.6 mm [4](70–80 % of the pump spot diameter) on the TD in order to support single transverse TEM_{00}-mode operation. In Fig. 3.3 a simple CW cavity consisting of two curved mirrors is shown. Such a cavity can be classified as a "convex–concave" type of cavity [25] and is known to support large beam sizes over the whole length. The TD and OC are end mirrors. Placing the TD as an end mirror has the advantage of affording the smallest possible accumulated wavefront distortions; on the other hand, it provides rather small gain. The resonator in Fig. 3.3 is dynamically stable. The mode size depends on the TD thermal lens (or pump power): as shown in Fig. 3.3, at near-zero pump power the TD has a dioptric power of −0.331/m and rather poor overlap of the fundamental mode with the pump spot. This results in a slightly distorted mode (see beam profile at ≈100 W pump power in Fig. 3.4). Furthermore, with an increase in pump power the beam profile changes to ideal fundamental shape with $M^2 = 1$ (as the overlap of the fundamental mode and pump spot is good enough) and slightly degrades ($M^2 = 1.1$) at pump powers around 400 W. This slight deterioration stems from exceeding 4 kW/cm^2 and thus causing weak beam wavefront distortions. The optical-to-optical efficiency is 45 % at 400 W pump, which is a typical value for CW TEM_{00} operation. This cavity can be modified into a configuration with the TD as folding mirror as shown in Fig. 3.5. The ABCD analysis of this cavity in Fig. 3.5 shows two stability zones arising as a function of the TD thermal lens, which agrees with the analysis presented in [29]. First, stability zone II is much narrower than that in Fig. 3.3. Thus the unpumped oscillator is located closer to the edge of stability zone II and pumping (changing the curvature of the TD) brings it closer to the other stability edge. The beam waist w_3 on the TD is nearly constant during this transition, which means good dynamic stability of the resonator and support of TEM_{00} operation. Moreover, it was derived in [29] that the width of the stability zone (I or II) scales as the inverse square of the mode radius on the the thermal lens $1/w_3^2$. This means that the resonators with large mode sizes on the TD are less dynamically stable.

3.3.3 Large Beam Sizes and Misalignment

As was described in Sect. 3.3.2 the increase of mode size comes at the expense of lower dynamic stability of the resonator. However, there is a second drawback related to the increased misalignment sensitivity. According to [29] the misalignment sensitivity can be defined by the parameter 1/C, where C is the absolute value of the

[4] defined as $1/e^2$ intensity drop

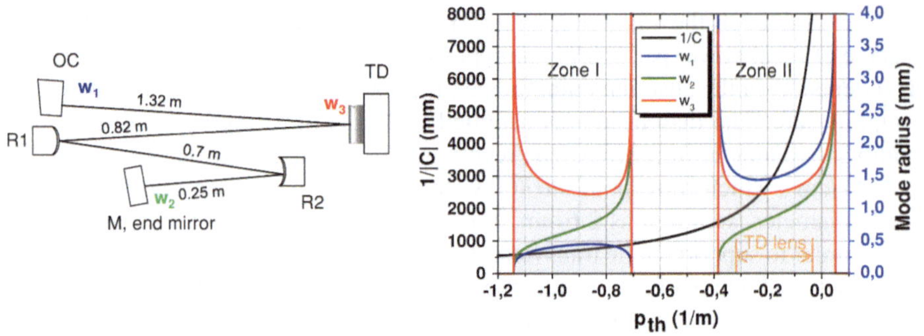

(a) (b)

Fig. 3.5 Modified version of the resonator in Fig. 3.3. The TD is placed as a folding cavity mirror. $R_{TD} \sim 6$m, R1, -4 m; R2, 4 m . Two stability zones arise as a function of the TD thermal lens (pump power)

equivalent focal length of the whole cavity. It is plain to see in Fig. 3.5b that the misalignment sensitivity is higher in zone II than in zone I. For our case the ratio of the misalignment sensitivities in the centres of the two stability zones totals 5, but the misalignment diverges when approaching the maximal pump powers (right orange line in Fig. 3.5) and getting closer to the edge of the zone II. The difference in misalignment sensitivity between the two zones can be simply explained by the larger beam sizes in stability zone II. Larger beam sizes on the mirrors unavoidably result in higher sensitivity to cavity alignment. Also the strong divergence at the stability edge of zone II is clearly related to the simultaneous divergence of the beam size on all mirrors. One could consider as the ultimate way of increasing the spot sizes on all mirrors to work at the right edge of zone II but at the cost of resonator mechanical stability. The resonator in Fig. 3.5 was operated in the whole range of pump powers and the difference in misalignment sensitivity was hardly noticeable.

3.4 Summary

During the last decade multi-mode CW TD lasers have proved to be a scalable and reliable tool for industrial applications. This was achieved with the Yb:YAG gain medium and until now this material remains unbeaten even in mode-locked operation. However, TEM_{00} operation especially in the mode-locked regime is more demanding with regard to resonator stability, strength of the TD thermal lens and the use of large beam areas. These constraints on the resonator and disk quality should be carefully analyzed in order to build a stable and reliable mode-locked oscillator.

References

1. A. Giesen, H. Hiigep, A. Voss, K. Wittig, U. Brauch, H. Opower, Scalable concept for diode-pumped high-power solid-state lasers. Appl. Phys. B **372**, 365–372 (1994)
2. G. Huber, C. Kränkel, K. Petermann, Solid-state lasers: status and future. J. Opt. Soc. Am. B **27**, B93–B105 (2010)
3. E. Sorokin, *Few-Cycle Laser Pulse Generation and Its Applications*, Solid-State Materials for Few-Cycle Pulse Generation and Amplification, Springer, Berlin, 2004), pp. 3–51
4. http://www.trumpf laser.com/produkte/festkoerperlaser/scheibenlaser/trudisk.html
5. http://www.ipgphotonics.com/apps_mat_multi_YLR.htm
6. M. Larionov, *Kontaktierung und Charakterisierung von Kristallen für Scheibenlaser* (Herbert Utz Verlag, München, 2009)
7. A. Giesen, J. Speiser, Fifteen years of work on thin-disk lasers: results and scaling laws. IEEE J. Quantum Electron. **13**(3), 598–609 (2007)
8. K. Beil, S.T. Fredrich-Thornton, F. Tellkamp, R. Peters, C. Kränkel, K. Petermann, G. Huber, Thermal and laser properties of Yb:LuAG for kW thin disk lasers. Opt. Express **18**(20), 20712–20722 (2010)
9. http://www.dausinger-giesen.de/products
10. R. Paschotta, J. Aus der Au, G. Spühler, S. Erhard, A. Giesen, U. Keller, Passive mode locking of thin-disk lasers: effects of spatial hole burning. Appl. Phys. B **72**, 267–278 (2001)
11. R. Paschotta, Power scalability as a precise concept for the evaluation of laser architectures. Open access paper on arXiv.org, (2007). http://www.arxiv.org/abs/0711.3987
12. C. Kränkel, Ytterbium-dotierte Borate und Vanadate mit großer Verstärkungsbandbreite als aktive Materialien im Scheibenlaser. PhD thesis, Universität Hamburg, (2008)
13. T. Südmeyer, C. Kränkel, C. Baer, O. Heckl, C. Saraceno, M. Golling, R. Peters, K. Petermann, G. Huber, U. Keller, High-power ultrafast thin disk laser oscillators and their potential for sub-100-femtosecond pulse generation. Appl. Phys. B **97**, 281–295 (2009)
14. R. Peters, Ytterbium-dotierte Sesquioxide als hocheffiziente Lasermaterialien. PhD thesis, Universität Hamburg, (2009)
15. B.L. Volodin, S.V. Dolgy, E.D. Melnik, E. Downs, J. Shaw, V.S. Ban, Wavelength stabilization and spectrum narrowing of high-power multimode laser diodes and arrays by use of volume Bragg gratings. Opt. Lett. **29**(16), 1891–1893 (2004)
16. O.H. Heckl, C. Siebert, D. Sutter, J. Kleinbauer, D. Bauer, Perfect precision in industrial micro machining. Laser Tech. J. **9**(2), 42–47 (2012)
17. private discussion with Kolja Beil
18. C.R.E. Baer, C. Kränkel, C.J. Saraceno, O.H. Heckl, M. Golling, R. Peters, K. Petermann, T. Südmeyer, G. Huber, U. Keller, Femtosecond thin-disk laser with 141 W of average power. Opt. Lett. **35**(13), 2302–2304 (2010)
19. D.S. Sumida, T.Y. Fan, *Emission spectra and fluorescence lifetime measurements of Yb:YAG as a function of temperature*, In advanced solid state lasers, (Optical Society of America, Washington, 1994), p. YL4
20. K. Petermann, G. Huber, L. Fornasiero, S. Kuch, E. Mix, V. Peters, S. Basun, Rare-earth-doped sesquioxides. J. Lumin. **87–89**, 973–975 (2000)
21. A. Killi, C. Stolzenburg, I. Zawischa, D. Sutter, J. Kleinbauer, S. Schad, R. Brockmann, S. Weiler, J. Neuhaus, S. Kalfhues, E. Mehner, D. Bauer, H. Schlueter, C. Schmitz, The broad applicability of the disk laser principle: from CW to ps, vol. 7193 (SPIE, Bellingham, 2009), p. 71931T
22. D. Bauer, I. Zawischa, D.H. Sutter, A. Killi, T. Dekorsy, Mode-locked Yb:YAG thin-disk oscillator with 41 μJ pulse energy at 145 W average infrared power and high power frequency conversion. Opt. Express **20**, 9698 (2012)
23. N. Hodgson, H. Weber, *Optical Resonators: Fundamentals, Advanced Concepts and Applications*, (Springer, Berlin, 1997)
24. O. Svelto, *Principles of Lasers*, 4th edn. (Springer, Heidelberg, 2007)

25. W. Koechner, *Solid-State Laser Engineering*, 6th edn. (Springer, Berlin, 1999)
26. S. Chetkin, G. Vdovin, Deformable mirror correction of a thermal lens induced in the active rod of a solid state laser. Opt. Commun. **100**, 159–165 (1993)
27. E. Schmid, J. Speiser, A. Giesen, *Characterisation of a Deformable Mirror for Compensation of the Thermal Lens in High Power Thin-Disk Lasers*, (Hamburg, Germany, 2010)
28. D. Hanna, C. Sawyers, M. Yuratich, Large volume TEM_{00} mode operation of Nd:YAG lasers. Opt. Commun. **37**(5), 359–362 (1981)
29. V. Magni, Multielement stable resonators containing a variable lens. J. Opt. Soc. Am. A **4**(10), 1962–1969 (1987)

Chapter 4
SESAM Mode-Locked Thin-Disk Oscillator

An introduction to the principle of SESAM and parameters such as saturation fluence, relaxation time, TPA-related effects and surface quality both before and after mounting is given in Sect. 4.1. A table summarizing the different SESAMs used in this work is also presented there. Section 4.2 describes thermal effects in dispersive mirrors, which are also crucial for the oscillator performance. This is followed by brief technical comments on the laser design. Experimental results and simulations are presented in Sect. 4.4. They show good agreement when TPA effects are included in the theoretical model. The SESAM damage issues are also discussed there. The chapter closes with concluding comments.

4.1 SESAM

SESAMs are a specific case of semiconductor saturable absorbers (SA) [1, 2]. SESAM appeared as a unification of a second cavity containing a multiple quantum well (MQW) saturable absorber [3, 4] in the resonant passive mode-locking technique [5, 6]. In the beginning the new monolithic construction representing a nonlinear mirror was called antiresonant Fabry Perot saturable absorber (A-FPSA) [7]. Since then the SESAM has become a widespread mode-locker in many oscillators, mainly those operated in the NIR spectral range: Nd:YLF [7], Yb:YAG [8], Cr:YAG [9], Ti:Sa [10] etc. Here, the SESAMs for 1 μm wavelength and >μJ pulse energies are emphasized.

4.1.1 SESAM Design

The acronym A-FPSA well describes the structural principle of a SESAM. It consists of the semiconductor Bragg mirror as a bottom reflector, typically comprising

O. Pronin, *Towards a Compact Thin-Disk-Based Femtosecond XUV Source*,
Springer Theses, DOI: 10.1007/978-3-319-01511-8_4,

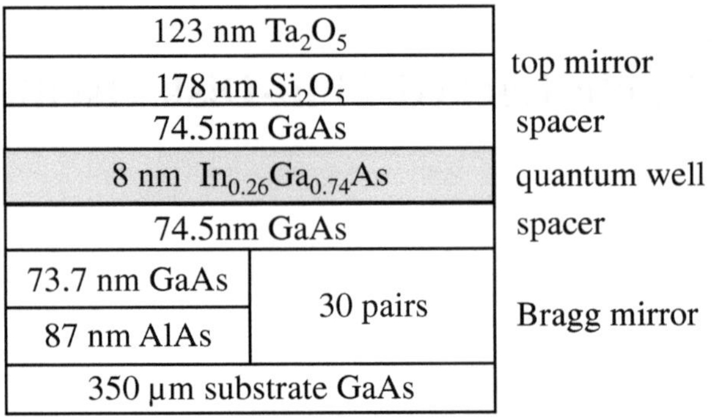

Fig. 4.1 Single QW SESAM designed for 1030 nm central wavelength

alternating GaAs/AlAs layers and a top reflector consisting of a dielectric coating or just the GaAs/air interface and in between these reflectors the semiconductor SA is embedded (Fig. 4.1), which can be a multiple QW, single-QW or bulk semiconductor. The SESAM structure shown in Fig. 4.1 is a particular case of mode-locking of a 1030 nm oscillator and is based on the works [6, 11]. In this SESAM saturable absorption occurs in the 8-nm-thick single QW of InGaAs. In the Fig. 4.1 the Bragg mirror consists of 30 alternating pairs of GaAs/AlAs in order to provide a sufficiently high reflectivity of 99.9 %.[1] Light penetrates into the structure and excites carriers from the valence to the conduction band. The process saturates for stronger light pulses because of band filling. The device can simply be used as a resonator end mirror with no serious constraints on resonator design.

4.1.2 SESAM Parameters and Properties

4.1.2.1 Relaxation Time

To date, both molecular beam epitaxy (MBE) and molecular chemical vapour deposition (MOCVD) growth technologies were used for SESAM manufacturing [12]. Ion implantation [13, 14] and low temperature (LT) growth [15] are known to reduce the relaxation lifetime in semiconductor saturable absorbers. The two methods induce different types of defects in the semiconductor crystal and decrease the carrier-capturing lifetime from ns down to ps time scales. MBE-LT-grown SESAM are most common. All samples used in this work, except for ANU1 (see Table 4.1) were LT-MBE grown.

[1] It is interesting to compare this with a standard high-reflective dielectric mirror used inside the oscillator. Such a mirror typically consists of SiO_2/Ta_2O_5 layers, in which case just 15 alternating pairs are enough to provide a reflectivity over 99.9 %.

Table 4.1 Summary of different SESAMs used in this work

SESAM	M4838	$M4838_{RTA}$	M4936	M4937	ANU1	Batop1	Batop2
ΔR,%	0.5	0.6	0.7	0.7	<2	0.7	0.5
ΔRns,%	1.5	–	0.5	0.5	–	0.3–0.5	0.3–0.5
F_{sat}, μJ/cm^2	160	30	40	50	–	30	120
F_2, mJ/cm^2	∞	∞	750	900	–	750	–
Top dielectric coating	No	No	No	No	No	Protective	Yes
Growth temperature T_g,°C	250	250	330	410	–	–	–
Fast relaxation time, fs	<1 ps	<1 ps	–	–	–	<1 ps	<0.5 ps
Long relaxation time, fs	<1 ps	<1 ps	>20 ps	>20 ps	–	<20 ps	–
Manufacture	WSI	WSI	WSI	WSI	ANU	Batop	Batop

ANU1 is MOCVD-grown and ion-implanted, all others LT-MBE. The data for Batop2 are specified according to the supplier

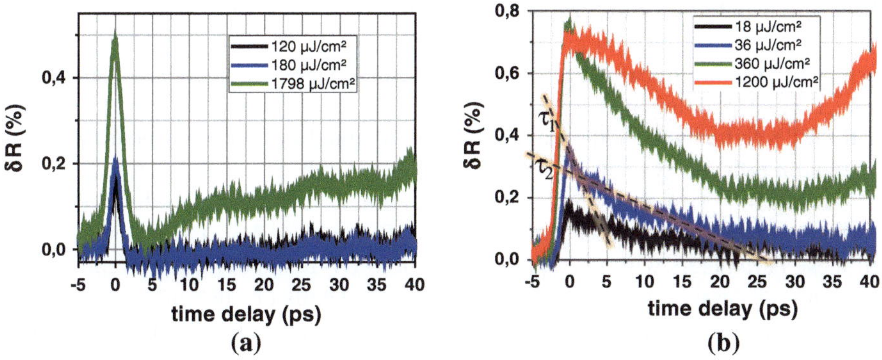

Fig. 4.2 Pump-probe measurement of the SESAM time response. **a** Sample M4838; **b** Sample Batop1. See Table 4.1

The whole SESAM structure is typically grown at around 550 °C; in contrast, the single InGaAs QW (see Fig. 4.1) is grown at lower temperatures, often starting at >200 °C. There are two recombination time scales: initial, rapid relaxation happens within the conduction band on a time scale <1 ps[2] (often called thermalization) and the second recombination from the conduction to valence band occurs on a time scale ≫1 ps. High defect concentrations lead to shorter relaxation times. In Fig. 4.2 a pump probe measurement[3] at different saturation fluences for samples grown at different temperatures is shown. The sample M4838 was grown at rather low temperature (250 °C) and the sample Batop1 at a temperature >350 °C.[4] We can see that the time response of M4838 is fairly short, <1 ps, and the measurement limited because the pulse duration in the pump-probe setup was ~1 ps. Such a short relaxation time

[2] Defined at 1/e of reflectivity drop.

[3] Measurements of dynamic SESAM reflectivity and time response were made by Farina Schättiger and Dominik Bauer [16].

[4] The exact value is not available.

can be explained by the large amount of defects; moreover this sample was not annealed. In the measurements in Fig. 4.2b for the Batop1 sample we can recognize two relaxation components: the first small spike, which appears at near zero time delay, and the second component, which relaxes at time delays approaching 40 ps. This behaviour is clearer for low saturation fluences (green and blue curves in Fig. 4.2b) and smears out at high incident fluences.[5]

The relaxation time of any SESAM is an important characteristic. Even though the pulses in a soliton mode-locked oscillator can be substantially shorter than the relaxation time of the mode-locker, it is always desirable to have the relaxation time as short as possible for the suppression of instabilities and generation of emission-bandwidth-limited pulses.

4.1.2.2 Absorbance and Saturation Fluence

In Fig. 4.3 the nonlinear reflectivity measurement shows a change of the SESAM reflectivity depending on the incident fluence. The fluence corresponding to a change of the reflectivity by 1/e of the maximal value is defined as saturation fluence F_{sat}. The maximum reflectivity is often called modulation depth ΔR. The difference between the absorption A of the sample and the saturated absorption (modulation depth) corresponds to non-saturable losses R_{ns}, i.e. $R_{ns} = A - \Delta R$. The measurement of all these parameters with good precision is non-trivial and demands a femtosecond oscillator at the wavelength of interest; for more details on such measurements see [17, 16]. Measurement of the static reflectivity can, however, be done relatively easily with a high-resolution spectrophotometer able to measure the absolute reflectivity. In Fig. 4.4 the measurements for the M4838 sample are shown. Upper curves correspond to the reflectivity of the calibrating dielectric mirror with known R > 99.98 % and the lower curves correspond to the reflectivity of the SESAM. The minimal size of the

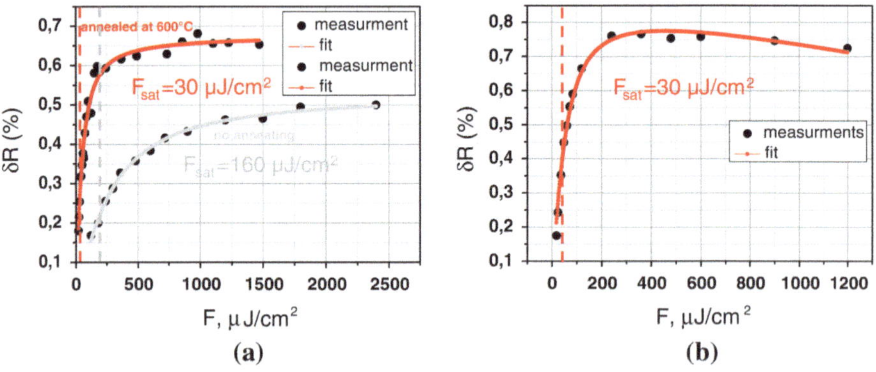

Fig. 4.3 Change of the SESAM reflectivity at different incident fluences

[5] The growing part in Fig. 4.2 can be caused by the scattered light at high incident fluences.

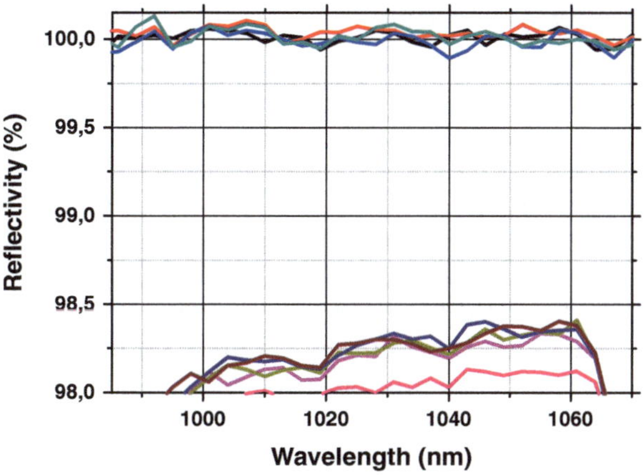

Fig. 4.4 Absoulute reflectivity measurements of the sample M4838. The *upper lines* correspond to the calibration measurement for the HR mirror. The *lower lines* correspond to the SESAM reflectivity (measured in a Perkin Elmer Lambda 950 spectrophotometer). Different lines correspond to different sample positions

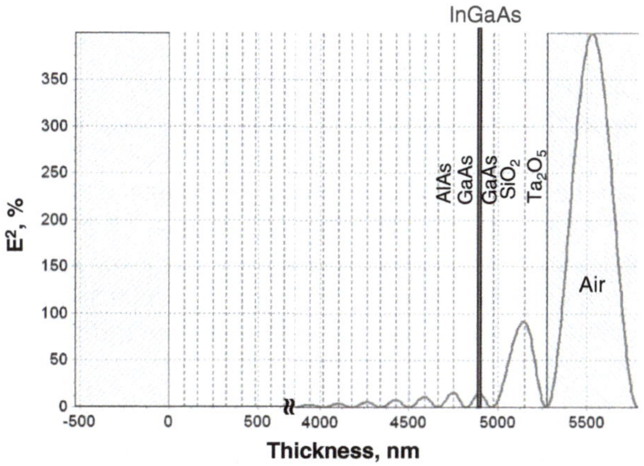

Fig. 4.5 Intensity field distribution in the SESAM structure (calculated in Optilayer software)

samples had to be $15 \times 15\,\text{mm}^2$. The accuracy of the measurements is 0.2–0.3 % and has statistical character: repositioning of the sample always results in slightly different reflectivity values (see Fig. 4.4). The absorption of the sample in Fig. 4.4 is $A \sim 1.5\,\%$. Subtraction of the saturable losses from the absorption yields $R_{ns} \sim 1\,\%$ non-saturable losses.

4.1.3 Two-Photon Absorption

The nonlinear saturable absorption plays a crucial role as mode-locking mechanism. The real absorption and excitation of the carriers is a key process for the SESAM mode-locked oscillator. As illustrated in Fig. 4.3b, the reflectivity not only saturates at some point, but even decreases at higher fluences. This effect is called inverse saturable absorption (ISA). Generally, it is a parasitic mechanism which does not allow an increase of the oscillator pulse energy as this tends to destabilize the regime via the onset of multi-pulsing. The ISA is directly related to TPA, which happens mainly in GaAs spacers and in end layers of Bragg mirrors (see Fig. 4.5). The light intensity is high in these parts of the structure and the TPA coefficient of GaAs $\beta_{GaAs} = 40\,\text{cm/GW}$ [18] is substantially higher than that for AlAs. Thus, the reduction of the field penetration depth into the structure and replacement of GaAs with a different material can shift TPA to higher fluences. The first measure can be accomplished by the deposition of a top dielectric coating, but at the expense of a reduced modulation depth. One can keep the modulation depth constant by placing more QWs into the structure, but at the expense of increased non-saturable losses [19, 20]. Replacing GaAs by any other material with lower TPA index has not been demonstrated so far at 1 μm wavelength.

Nevertheless, it can be considered advantageous to have TPA for stabilizing the regime against the onset of Q-switched ML instabilities [21, 22], see Sect. 4.4. These instabilities are typically well suppressed in high-energy oscillators which, however, is due to high saturation of the gain medium and absorber.

TPA can easily be seen in the experiment by looking at the beam spot on the SESAM and monitoring the green TPA light. Sometimes even the onset of multiple pulses can be indicated as a change of brightness in the green spot on the SESAM.

Later on, in Sect. 4.4.7 the influence of TPA on the maximal oscillator energy and minimal pulse duration will be demonstrated both theoretically and experimentally.

4.1.4 SESAM Surface and Mounting Techniques

SESAMs have saturable and non-saturable losses. Both types of losses contribute to heating of the device. As one can see in Table 4.1, non-saturable losses are approximately 30–40 % of the overall SESAM absorption and for the 250 °C-grown M4838 sample it is worse. For instance, the specified absorption of Batop2 is 1 % with a modulation depth of $\Delta R = 0.5\,\%$, resulting in non-bleachable losses of $R_{ns} = 0.5\,\%$. Heating is an undesirable effect because it causes a thermal lens and may reduce the damage threshold of the semiconductor chip. Therefore it is important to ensure effective heat removal. This can be done analogous to the TD mounting [23]. The simplest technique which can be applied in-house is an epoxy mounting. The typical thermal conductivity of epoxies is in the range of 1–2 W/(mK); in rare cases a conductivity of 11 W/(mK)[6] can be achieved. These values have to be compared with the thermal conductivity of a GaAs wafer $\kappa_{GaAs} = 55$ W/(mK) in order to understand that the main thermal resistance in the heat flow from the SESAM structure to metal heat sink is caused by the epoxy layer, which in this case has to be as thin as possible. Even though the epoxy mounting can be done in-house, achieving thin bonding layers <5 μm in combination with good uniform contacting is very difficult. Such mounting is technically difficult and has to be done under pressure with fast epoxy hardening [23].

The soldering technique can be considered as an alternative to epoxy mounting. Preliminary metallization of the back surface of the chip has to be done. The soldering alloy has a thermal conductivity as high as the heat sink and, being relatively thick, smoothes surface defects mainly originating from the copper heat sink. Both epoxy-mounted and soldered SESAMs were used in this work. Different SESAM samples were measured to study the surface quality change after mounting to a heat sink. In Fig. 4.6a–c the interferometric surface measurement[7] of a free standing SESAM sample is shown. From Fig. 4.6a it can be seen that the surface is of convex shape, having a flatness deviation of 1.4 μm, which corresponds to a radius of curvature of R = 3.2 m. Here, we do not take into account the deviations in the corners of the sample which arise from the rectangular shape of the sample. The SESAM has the astigmatic shape shown in Fig. 4.6b after subtraction of the spherical part. Figure 4.6c shows the result of subtracting the spherical and astigmatic parts, indicating a

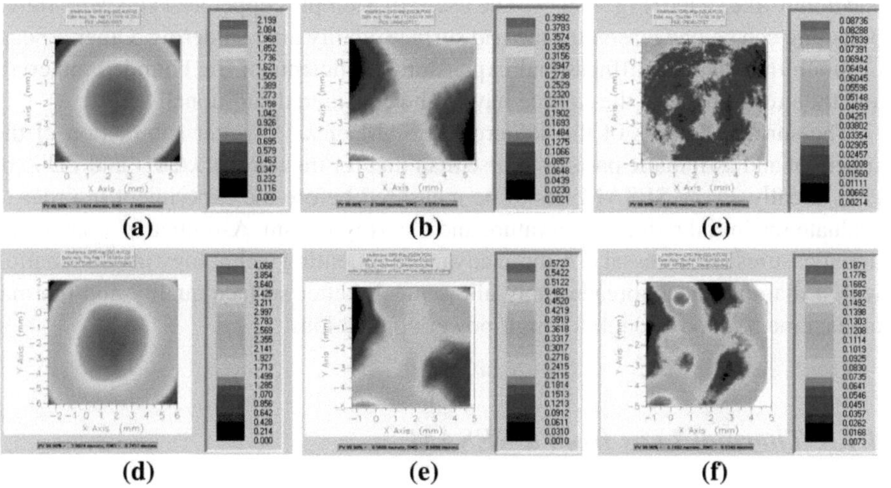

Fig. 4.6 Interferometric measurement of M4838 before soldering (**a**–**c**) and after soldering (**d**–**f**); **b**, **e**: spherical part is subtracted; **c**, **f**: spherical and astigmatic parts are subtracted. Similar measurements were performed with soldered SESAMs from Batop GmbH and showed similar surface deviations

[6] This is Epo-Tek EK 1000 from Epoxy Technology.

[7] The measurements in Fig. 4.6 were done by Mikhail Larionov.

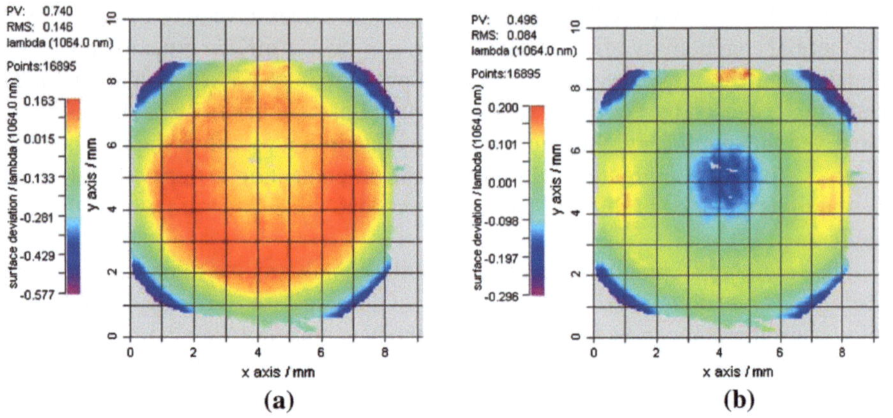

Fig. 4.7 Interferometric measurement of the Batop1 epoxy mounted sample. The spherical part is subtracted in case (**b**)

flatness deviation of ~80 nm. Soldering increases the stress and leads to a stronger convex curvature R = 1.7 m. It also increases astigmatism and impairs the surface quality. The maximum flatness deviation after soldering is ~190 nm (see Fig. 4.6f) for the astigmatism-subtracted case. Moreover, other measurements show variations of the surface curvature and quality of different samples cleaved from the same wafer. The thick-layer epoxy-mounted chips are of comparable quality to the soldered ones and donut-like shape (Fig. 4.7).[8,9] The usage of a simple copper heat sink with rough polishing can be one reason for the bad surface quality after soldering. Other aspects of mounting such as stiffness and expansion coefficients of the bounded materials and the initial shape of the sample have to be considered additionally.

The surface flatness of the soldered SESAM of $\lambda/5$[10] (after subtraction of the spherical and astigmatic parts) is worse than that for the mirrors $\lambda/20$[11] and TD $\lambda/10$ [24]. Ideally, every SESAM has to be measured before the experiment in order to evaluate the initial radius of curvature and the astigmatism. Astigmatism compensation measures are necessary. A more advanced mounting technology in combination with a diamond heat spreader may allow much better surface quality and thermal management, good enough for high-power applications.

4.1.5 Thermal Lens from a SESAM

Even though the SESAM can be properly mounted to a heat sink to improve thermal management, it does exhibit thermal effects which can be characterized by an

---

8 More systematic investigation of the surface of epoxy-mounted chips was not performed.

9 The measurements in Fig. 4.7 were done by Dominik Bauer.

10 λ is 1000 nm

11 High-quality Layertec mirrors for instance.

interferometer embedded into the cavity analogously to similar investigations performed for the TD [23]. This complex approach was not carried out due to its complexity and a lack of time. Generally, it was noticed that the epoxy-mounted SESAMs have worse thermal management than the soldered ones. Therefore, only the soldered SESAMs were used for high-power experiments with intracavity average power levels >500 W. Measures to compensate the thermal lens were performed by means of the telescopic arrangement in Fig. 4.9. It consists of two mirrors with *R1* and *R2* (for most experiments R1 = R2 = 0.3 m). The change of the separation distance *d* between them leads to compensation of the thermal lens from the SESAM. This telescopic arrangement was also used to pre-compensate the initial radius of curvature of the SESAM, which varies from sample to sample. In the light of cavity sensitivity to a thermal lens [25] and the power scaling concept [26], the thermal lens induced by the SESAM in conjunction with the surface quality is of crucial importance for stable operation and is one of the limitations in increasing the average output power.

4.1.6 SESAMs in This Work

Different SESAMs used in the experiments are now summarized.

4.2 Thermal Effects in Dispersive Mirrors

Not only the TD and SESAM exhibit a thermal lens, high-dispersive (HD) mirrors are also influenced by thermal effects. It can be observed in a simple CW cavity as in Fig. 3.3 by replacing some of the HR mirrors with HD mirrors. In the worst case at intracavity power levels >1 kW this may result in a different mode profile, slightly distorted mode and beam-pointing drifts, indicating a slow and time-dependent misalignment of the cavity. Different HD mirrors were experimentally studied and the presence of such effects was verified. At GDD values typically higher than 1000 fs^2 power-dependent effects start to appear. These power-dependent effects cannot be identified completely as purely thermal ones. In Fig. 4.8 the surface temperature of different HD mirrors placed inside the cavity was measured in CW operation with ~0.8 kW intracavity power. The measurements were done with an infrared camera (FLIR SC305) and indicate a certain temperature rise. It can be seen in Fig. 4.8a–c that samples HD64 ($-1000\,fs^2$), HD73 ($-3000\,fs^2$) and Layertec ($-1000\,fs^2$) exhibit a similar temperature rise of 10–20 K. These mirrors show no noticeable temperature-dependent effects during oscillator operation. The mirror HD 105 ($-3000\,fs^2$) has a pronounced temperature rise of 50 K and causes the mode and the oscillator stability to deteriorate at high intracavity power >1 kW. Still it is difficult to relate these observations to the mirror design (light penetration and resonant effects in the structure) [27, 28], overall layers thickness, amount of GDD and material properties. More

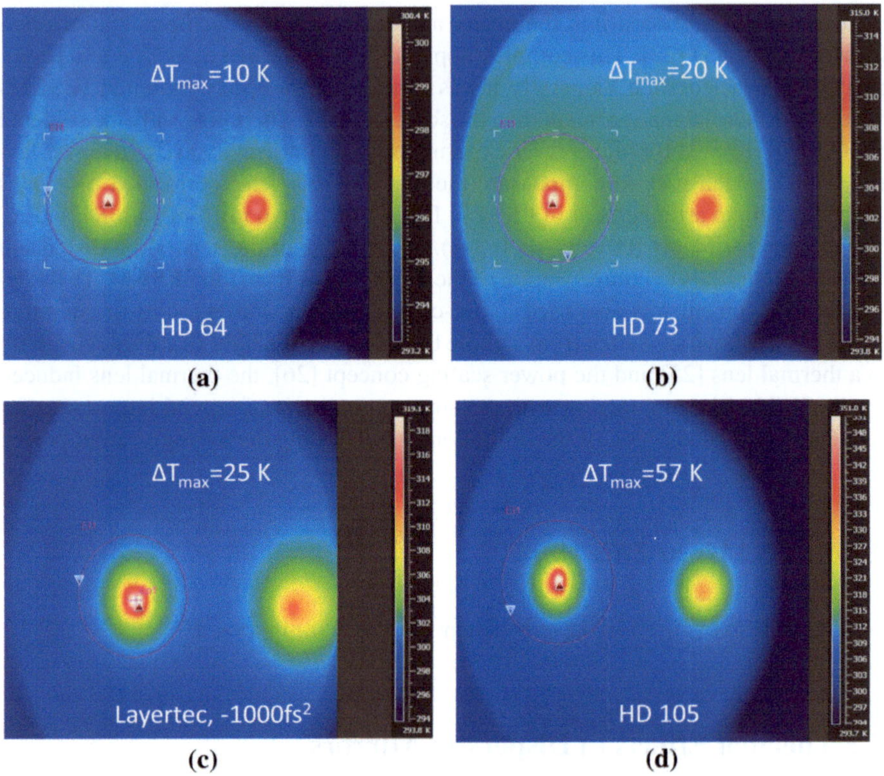

Fig. 4.8 Surface temperature of different HD mirrors measured with infrared camera (FLIR SC305) at CW intracavity power of ∼0.8 kW

systematic studies and experiments have to be done to evaluate the main reason for the high-power-dependent behaviour of the HD mirrors.

Water cooling of mirrors and the use of crystalline substrates such as sapphire or YAG, which have better thermal conductivity than fused silica, can be considered to improve the power-dependent behaviour of the HD mirrors. Ultimately, the implementation of a power-scalable geometry similar to the TD may become necessary for multi-kW intracavity power levels.

4.3 Technical Aspects

The overall laser performance depends on many technical issues. High-intracavity-power TD oscillators are sensitive to dust. Dust particles, when they appear on the mirror surface, may cause damage during laser operation or, at least, will be burned in. In order to avoid this the laser system has to be operated together with a flow box or placed in an air-tight housing with small over-pressure. In both cases the housing has to dampen external acoustic noise and be tight enough to prevent air turbulence. There should be no overheated or overcooled components inside the cavity in order to prevent air flow between them. Water-cooled components normally induce vibrations. These vibrations may originate from chiller vibrations and can easily be decoupled by means of longer and softer water tubes. The flow fluctuations can be damped by special flow dampeners. The temperature stability in the lab as well as the humidity may also be issues. As was found in some experiments [24, 29, 30], the water-cooling concept of the TD is itself a source of acoustic noise. Thus the flow should be reduced to a minimum in order to prevent these vibrations. In this work a water flow on the TD of 0.5–0.8 l/min was used. The simple and inexpensive housing construction for the laser is described in Appendix B. The aspects described below are of great importance when more precise experiments such as repetition rate synchronization [31] and CEP stabilization [32] have to be done with the oscillator. The PhD student has to remember that saving time on such technical things may afterwards result in useless oscillator performance (the author has proved this in an experiment on himself).

4.4 Mode-Locking Experiments

In this section, a description of consequent efforts on oscillator mode-locking with different SESAMs is given. Every subsection will be devoted to the specific SESAM type and corresponding oscillator parameters. Our experimental setup, sketched in Fig. 4.9, includes a 220-μm thin-wedged Yb:YAG disk of 7 % doping (TD157, see Table 3.1). A disk head is aligned for 24 passes of the pump beam through the gain medium with a pump spot diameter of 3.2 mm. The Yb:YAG TD is used as one of the folding mirrors in a standing-wave cavity pumped by fibre-coupled diodes centred at a wavelength of 940 nm. The cavity is a convex–concave type designed for providing large mode sizes over its entire length, in order to minimize nonlinear propagation effects in air and also the risk of damage to optical components. As we can see in Fig. 4.9, the average mode radius inside the cavity is >1 mm, and the beam radius on the disk is ~ 1.2 mm and 0.9 mm on the end mirror incorporating a SESAM. The end part of the cavity contains a telescope for compensating a thermal lens induced by the SESAM and for scaling the beam radius on it [33].

4.4.1 Q-Switched Mode-Locking and Self-Q-Switching

Before going into the experimental results on mode-locking it is necessary to make a comment on two terms: Q-switched mode-locking and self-Q-switching.[12] Both of them apply to Q-switched-like instabilities and are very often used as syn-

[12] Sometimes the term self-Q-switching also means passive Q-switching.

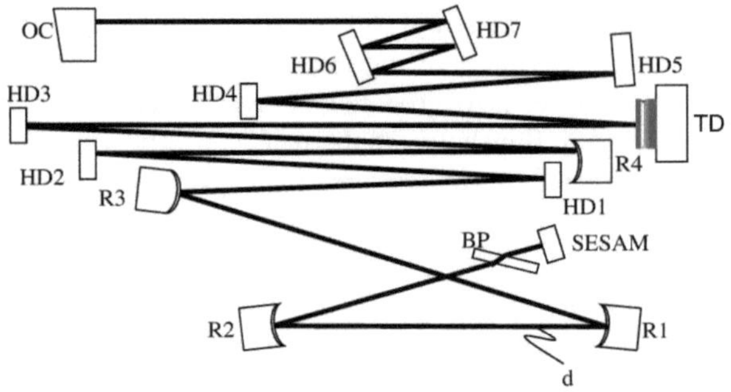

Fig. 4.9 Schematic of the SESAM mode-locked Yb:YAG disk oscillator. All flat mirrors are HD mirrors with $-1000\,\text{fs}^2$ and $-3000\,\text{fs}^2$ per bounce; OC, an output coupler with 5.5 or 9.3 % transmission; *R1*, *R2* concave mirrors with 0.3 m radius of curvature ; $R3 = -4\,\text{m}$; $R4 = 4\,\text{m}$; BP, 1-mm-thick fused-silica Brewster plate. The pulse repetition rate is 40 MHz. The oscillator is placed inside a box of size $1 \times 0.4\,\text{m}^2$

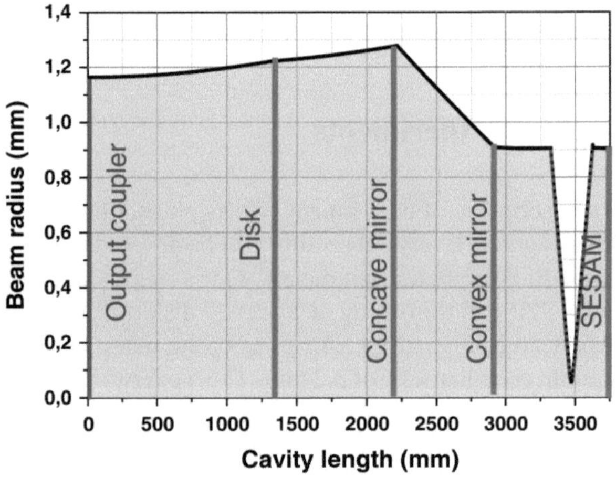

Fig. 4.10 Mode radius vs cavity length. Calculated for the concave TD with $R_{TD} = 25\,\text{m}$

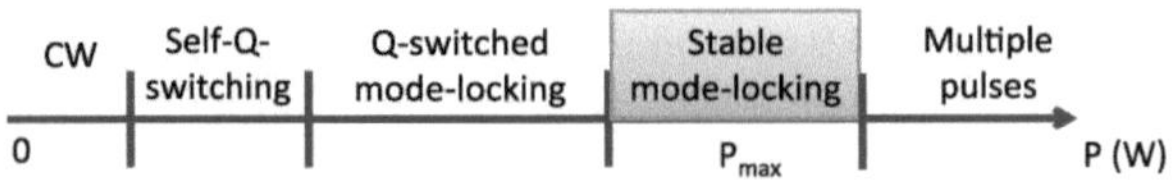

Fig. 4.11 Picture illustrates different instabilities arising during increasing of the oscillator pump power. Self-Q-switching, when it happens, causes SESAM damage

onyms in the literature. A slight difference is made between them in this work. During the work on the SESAM mode-locked oscillator several situations relating to Q-switched instabilities were observed. Often the oscillator started to strongly Q-switch at low output powers and a further increase of pump power resulted in damage to the SESAM. Q-switching was strongly chaotic, typically accompanied by high-frequency acoustic noise. This was in contrast to the typical description of Q-switched mode-locking as a slowly modulated mode-locked pulse train. For clarity, this strong regime with early onset is hereafter called "self-Q-switchin". Moreover, the author is not aware of any publication mentioning this regime as a strongly limiting factor Fig. 4.10. Here, the term "self-Q-switching" should not be confused with passive Q-switching when there is no mode-locking Fig. 4.11.

4.4.2 ML with M4838 and M4838$_{RTA}$

As can be seen in Table 4.1, the single QW of the M4838 sample was grown at a low temperature of $T_g = 250\,°\mathrm{C}$ and had no post-annealing. Afterwards the wafer was cleaved in to $8\times 8\,\mathrm{mm}^2$ size samples and epoxy-mounted to the copper heat sink. Insertion of this SESAM into the cavity resulted in stable mode-locked operation with an output power of 45 W with a 9 % OC. The oscillator was not self-starting and had to be perturbed to initiate mode-locking. Negligible self-Q-switching was observed during increasing of the pump power up to P_{max} <200 W. The corresponding spectrum and microwave beat signal are shown in Fig. 4.12. The net cavity GDD was $\sim -14000\,\mathrm{fs}^2$. The thermal effects in the chip were clearly a limiting factor for further power increase. Thus, both annealing of this sample to decrease non-saturable losses and soldering to the copper heat sink to improve heat dissipation were performed. This sample was marked as M4838$_{RTA}$ (see Table 4.1). However, an increase of the pump power with this sample caused self-Q-switching already at output powers <10 W. A further increase of the pump power induced stronger self-Q-switching and, finally, damage. The self-Q-switching threshold varied negligibly, depending on the amount of GDD, cavity alignment and the spot size on the SESAM. Therefore this mode of operation could not be avoided.

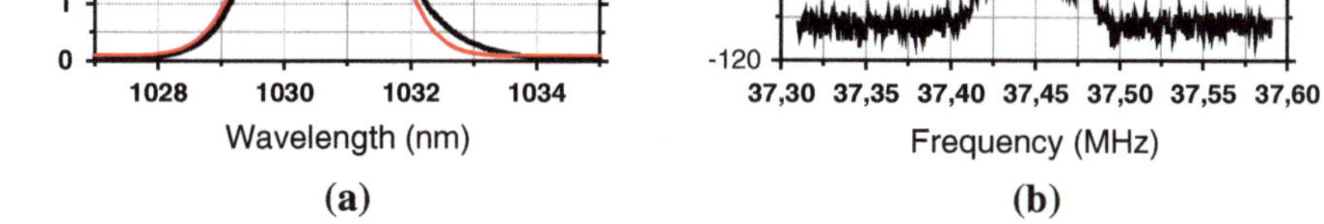

Fig. 4.12 **a** Optical spectrum and mode profile at 45 W output power; **b** microwave beat signal showing 50 dBm sideband suppression

4.4.3 ML with M4936 and M4937

The samples M4936 and M4937 had single InGaAs QW grown at 330 and 410 °C, respectively. The experiments with these samples always resulted in damage to them through self-Q-switching. The deposition of a top dielectric coating on these samples improved the situation and no self-Q-switching could be observed. The oscillator with samples having two layers of Ta_2O_5 (128 nm) and SiO_2 (178 nm) could be mode-locked. However, the dielectric coating caused a decrease of the modulation depth and prevented us from reaching higher pulse energy than that described in the previous experiment.

4.4.4 ML with ANU1

The ANU1 sample was MOCVD-grown and had three QWs, placed in the one antinode of the field distribution. Different samples were available depending on the oxygen implantation dose 1.8–2.4×10^{14} 1/cm^2. For more details on this SESAM type see [12]. The performance of ANU1, epoxy-mounted to a heat sink, was poor due to the thermal problems, indicated by strong output beam deterioration. Moreover, it also exhibited self-Q-switching. The deposition of a top dielectric coating (two layers Ta_2O_5, SiO_2) eliminated self-Q-switching, yet the problems with beam deterioration remained the same. No further experiments were done with this type of SESAM.

4.4.5 ML with Batop1

Additionally, many other samples from Batop GmbH were tried out from four different wafers with modulation depths <1 % and fast relaxation component <1 ps. These samples were damaged mainly due to self-Q-switching. A couple of them, however, resulted in stable oscillator performance. Batop1 is one of these chips. A remarkable feature was the strong irreproducibility of the self-Q-switching threshold in day-to-day operation. Damage because of self-Q-switching and stable operation was occurring in the ratio 50/50. No remarkable output power increase was achieved in relation to the previous results. It is interesting to note that Batop1 was designed for a slightly longer wavelength 1040 nm and thus had a saturation fluence of 30 μJ/cm^2 at 1030 nm, which differs from that of 90 μJ/cm^2 specified for 1040 nm.

Even samples from the same wafer, supposedly identical, behaved differently, thus evidencing poor reproducibility.

4.4.6 ML with Batop2

The most valuable results on SESAM mode-locking were achieved with the sample Batop2. It has specifications similar to Batop1, but with a top dielectric coating

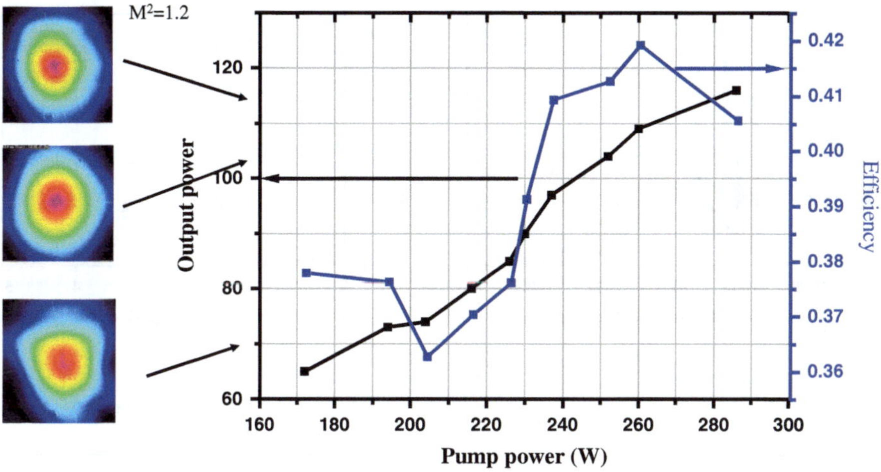

Fig. 4.13 Beam profile, output power and efficiency vs pump power

(Table 4.1). Unfortunately, details about coating and structure are not available. The oscillator running with Batop2 showed no tendency to self-Q-switching and was often self-starting. This self-starting brought the oscillator into the Q-switched mode-locked regime, and further increase of the pump power resulted in stably mode-locked operation. The stable range for pumping was rather narrow with powers between 210–230 W (see Fig. 4.13). The system has indeed a strong hysteresis: once mode-locked at 210 W the system keeps running stably at a decreased pump power down to 170 W, where it starts to Q-switch. A further increase in output power was accomplished by changing the distance *d* (Fig. 4.9) in proportion to the growing pump power. In order to reach higher output energies the GDD was increased up to ~-30.000 fs^2 together with the pump power. Nevertheless, the soliton could not be stabilized by the rather low modulation depth <0.5 % of Batop2 and as a consequence additional CW peaks (in Fig. 4.14) appeared at higher output power. Integration of the area under these peaks yields approximately 20–30 % of the overall power. TPA could also be a reason for this instability. An attempt to increase the mode diameter on the SESAM from 1.8 to 2.5 mm by replacing mirrors *R1* and *R2* with different ones caused a very strong sensitivity of the oscillator to the TD/SESAM thermal lens. A further output power increase >120 W was pointless.

4.4.7 Experiment and Simulations

Analytical calculations[13] were performed in order to understand the main causes prohibiting a further energy increase. A variational approach [34] to the generalized complex nonlinear Ginzburg-Landau equation (see Sect. 2.2) and experimental

[13] By Vladimir Kalashnikov.

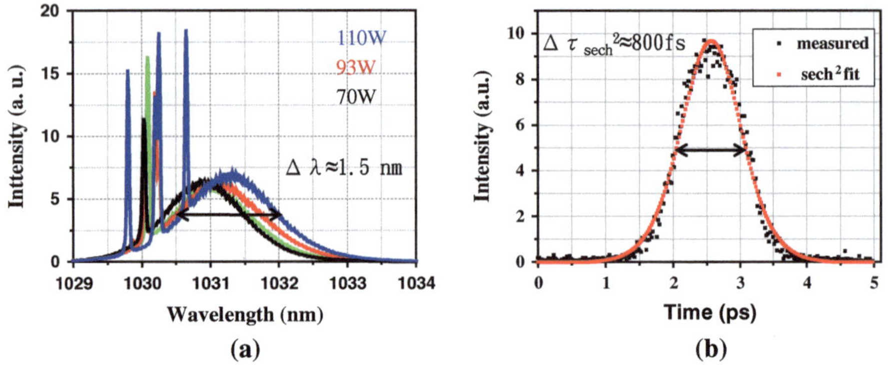

Fig. 4.14 Oscillator characteristics operating with Batop2 sample. **a** Optical spectra; **b** AC trace

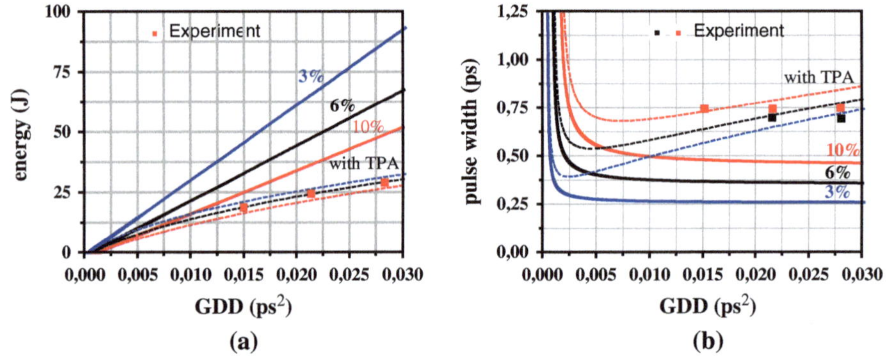

Fig. 4.15 **a** Intracavity pulse energy vs GDD; **b** pulse duration vs GDD. *Solid curves* correspond to the calculations with TPA, dashed to those without. *Red* and *black squares* represent experimental data at 10 % and 6 % OC transmission, respectively

results are compared in Fig. 4.15.[14] Unsaturable loss and saturable gain, spectral filtering, GDD, SPM induced by air, the plate and the laser disk, and perfectly saturable losses in the SESAM were taken into account in these calculations. The TPA parameter was d = 0.00025, where $d = \theta/\zeta$, θ is the TPA coefficient (in W^{-1}) and ζ is the inverse power of the loss saturation. Two cases are considered: with influence of TPA in the SESAM and without it. The parameters for the simulations were chosen to resemble Batop2 and respective results. Solid curves correspond to the calculations with TPA, dashed to those without. The transmission of the output coupler varied between the three values 3, 6 and 10 %. Red and black squares represent experimental data at 10 % and 6 % OC transmission, respectively. The curves show marginal stability of the mode-locked regime; in other words, the area below the curves corresponds to the unstable region. The analytical results including a TPA effect fit much better to the values observed in the experiment. It is obvious that TPA

[14] The calculations were done by Vladimir Kalashnikov.

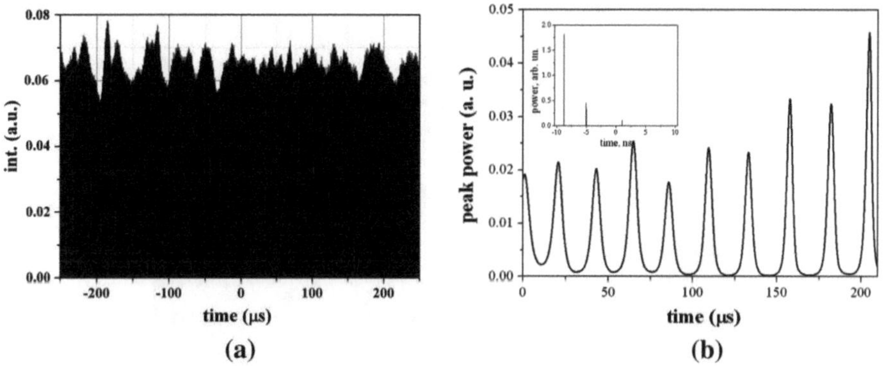

Fig. 4.16 **a** Chaotic modulation of the pulse train observed in the experiment; **b** numerical simulation qualitatively reproducing this regime

strongly reduces the maximum attainable energy of the system (compare solid and dashed curves in Fig. 4.15a) and increases the minimal attainable output pulse duration (compare solid and dashed curves in Fig. 4.15b). Therefore the most critical limit for further energy scaling can be identified to be TPA in the SESAM. As was already mentioned (Sect. 3.3), the resonator stability zone shrinks at larger spot diameters (>1.8 mm) on the SESAM and prevents pumping at higher powers, because of the thermal lensing in TD and SESAM. Thus reduction of TPA via enlarging the spot size on the SESAM was not possible.

4.4.7.1 Chaotic Q-Switching

An interesting regime was observed with the Batop2 sample and GDD $= -13.000\,\mathrm{fs}^2$ when trying to reach intracavity energies higher than $\sim 14\,\mu\mathrm{J}$. The onset of multiple pulses accompanied by chaotic-like Q-switched mode-locking occurred. This chaotic regime can be a limiting factor in energy increase and is a very interesting subject for theoretical investigations in itself. Numerical simulations performed for this case revealed similar oscillator behaviour and showed that closely (∼ps) and far-separated solitons interact with each other via dynamic gain saturation. Evidently, the latter had to be included in the numerical simulations of the complex nonlinear Ginzburg-Landau equation.

In conclusion, experimental data and theoretical calculations agree well with each other when TPA is included in the model.

4.4.8 SESAM Damage

All optical intracavity components are subject to damage. The most sensitive elements are SESAMs and HD mirrors. Two regimes of the oscillator should be

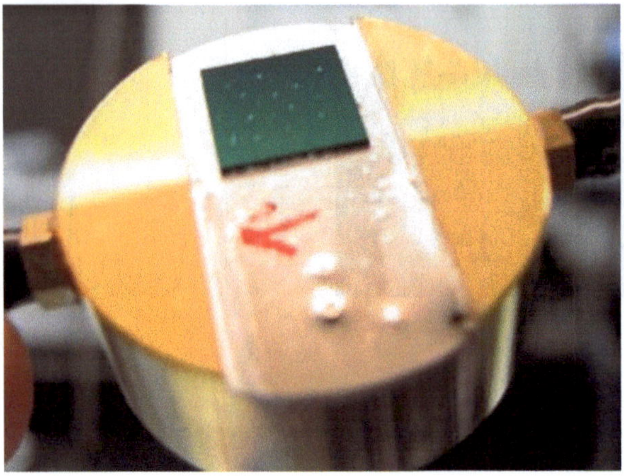

Fig. 4.17 The soldered sample Batop2 (from Batop GmbH) peppered with plenty of damaged spots. The *green color* of the chip is due to the dielectric coating

distinguished: a pulse build-up (starting regime) and a stable mode-locked regime. Mostly, all damage occurs during the pulse build-up and is related to the instabilities accompanying this process and to the correspondingly high-peak intensities. Sometimes, damage may occur when stable operation of the oscillator is perturbed by dropping something on the optical table or misaligning the cavity. Therefore, it is of crucial importance to optimize the starting regime of the oscillator and keep the oscillator away from external perturbations. The damage threshold should be addressed with a coefficient F_2 characterizing the roll-over in nonlinear SESAM reflectivity and the pulse duration at which this coefficient was measured. Otherwise it is not clear which limitation will be most crucial: the onset of multiple pulses or the SESAM damage. The damage threshold of the SESAM strongly depends on the design and is subject to a certain trade-off. These issues together with SESAM degradation were studied in detail in [20].

No degradation effects were noticed in this work as the oscillator was not operated for a long time with one sample Fig. 4.16.

It was found in our work that most SESAMs were damaged via self-Q-switching without the attaining an optimal operation point. This turned out to be the most serious limitation to using the SESAMs Fig. 4.17.

4.5 Summary

Many experiments were conducted with different SESAMs during this work. The semiconductor chips showed limited performance as regards surface quality, damage due to the self-Q-switching, TPA, heating and corresponding thermal effects. Simultaneous high-power (>1 kW intracavity), high-energy (>20 μJ) and short-pulse (<300 fs) operation of the oscillator is limited due to the above-listed shortcomings. Serious technological improvements are needed to overcome these shortcomings.

Nevertheless, the SESAM can serve as a starter [35] for KLM. The design of the starter SESAM is much more flexible and there are fewer constraints on it.

References

1. M. Islam, E. Sunderman, C. Soccolich, I. Bar-Joseph, N. Sauer, T. Chang, B. Miller, Color center lasers passively mode locked by quantum wells. IEEE J. Quantum Electron. (1989)
2. A.F. Gibson, M.F. Kimmitt, B. Norris, Generation of bandwidth-limited pulses from a TEA CO_2 laser using p-type germanium. Appl. Phys. Lett. **24**(7), 306–307 (1974)
3. C.L. Cesar, M.N. Islam, C.E. Soccolich, R.D. Feldman, R.F. Austin, K.R. German, Femtosecond KCl:Li and RbCl:Li color-center lasers near 2.8 μm with a HgCdTe multiple-quantum-well saturable absorber. Opt. Lett. **15**(20), 1147–1149 (1990)
4. C.E. Soccolich, M.N. Islam, M.G. Young, B.I. Miller, Bulk semiconductor saturable absorber for a NaCl color center laser. Appl. Phys. Lett. **56**(22), 2177–2179 (1990)
5. U. Keller, W. Knox, G. 'tHooft, Ultrafast solid-state mode-locked lasers using resonant nonlinearities. IEEE J. Quantum Electron. **28**(10), 2123–2133 (1992)
6. U. Keller, K. Weingarten, F. Kärtner, D. Kopf, B. Braun, I. Jung, R. Fluck, C. Honninger, N. Matuschek, J. Aus der Au, Semiconductor saturable absorber mirrors (SESAM's) for femtosecond to nanosecond pulse generation in solid-state lasers. IEEE J. Quantum Electron. **2**(3), 435–453 (1996)
7. U. Keller, D.A.B. Miller, G.D. Boyd, T.H. Chiu, J.F. Ferguson, M.T. Asom, Solid-state low-loss intracavity saturable absorber for Nd:YLF lasers: an antiresonant semiconductor Fabry-Perot saturable absorber. Opt. Lett. **17**(7), 505–507 (1992)
8. J.A. der Au, G.J. Spühler, T. Südmeyer, R. Paschotta, R. Hövel, M. Moser, S. Erhard, M. Karszewski, A. Giesen, U. Keller, 16.2-W average power from a diode-pumped femtosecond Yb:YAG thin disk laser. Opt. Lett. **25**(11), 859–861 (2000)
9. B.C. Collings, J.B. Stark, S. Tsuda, W.H. Knox, J.E. Cunningham, W.Y. Jan, R. Pathak, K. Bergman, Saturable Bragg reflector self-starting passive mode locking of a Cr^{4+}:YAG laser pumped with a diode-pumped Nd:YVO_4 laser. Opt. Lett. **21**(15), 1171–1173 (1996)
10. L. Brovelli, I. Jung, D. Kopf, M. Kamp, M. Moser, F. Kärtner, U. Keller, Self-starting soliton modelocked Ti-sapphire laser using a thin semiconductor saturable absorber. Electron. Lett. **31**(4), 287–289 (1995)
11. S.V. Marchese, C.R. Baer, A.G. Engqvist, S. Hashimoto, D.J. Maas, M. Golling, T. Südmeyer, U. Keller, Femtosecond thin disk laser oscillator with pulse energy beyond the 10-microjoule level. Opt. Express **16**(9), 6397–6407 (2008)
12. M.J. Lederer, V. Kolev, B. Luther-Davies, H.H. Tan, C. Jagadish, Ion-implanted InGaAs single quantum well semiconductor saturable absorber mirrors for passive mode-locking. J. Phys. D **34**(16), 2455 (2001)
13. M. Lambsdorff, J. Kuhl, J. Rosenzweig, A. Axmann, J. Schneider, Subpicosecond carrier lifetimes in radiation-damaged GaAs. Appl. Phys. Lett. **58**(17), 1881–1883 (1991)
14. M.J. Lederer, B. Luther-Davies, H.H. Tan, C. Jagadish, M. Haiml, U. Siegner, U. Keller, Nonlinear optical absorption and temporal response of arsenic- and oxygen-implanted GaAs. Appl. Phys. Lett. **74**(14), 1993–1995 (1999)
15. E.S. Harmon, M.R. Melloch, J.M. Woodall, D.D. Nolte, N. Otsuka, C.L. Chang, Carrier lifetime versus anneal in low temperature growth GaAs. Appl. Phys. Lett. **63**(16), 2248–2250 (1993)

16. F. Schättiger, D. Bauer, J. Demsar, T. Dekorsy, J. Kleinbauer, D. Sutter, J. Puustinen, M. Guina, Characterization of InGaAs and InGaAsN semiconductor saturable absorber mirrors for high-power mode-locked thin-disk lasers. Appl. Phys. B **106**, 605–612 (2012)
17. M. Haiml, R. Grange, U. Keller, Optical characterization of semiconductor saturable absorbers. Appl. Phys. B **79**, 331–339 (2004)
18. A. Saïssy, A. Azema, J. Botineau, F. Gires, Absolute measurement of the 1.06 μm two-photon absorption coefficient in GaAs. Appl. Phys. A **15**, 99–102 (1978)
19. C.J. Saraceno, O.H. Heckl, C.R.E. Baer, M. Golling, T. Südmeyer, K. Beil, C. Kränkel, K. Petermann, G. Huber, U. Keller, Sesams for high-power femtosecond modelocking: power scaling of an Yb:$LuScO_3$ thin disk laser to 23 W and 235 fs. Opt. Express **19**(21), 20288–20300 (2011)
20. C.J. Saraceno, C. Schriber, M. Mangold, M. Hoffmann, O.H. Heckl, C.R.E. Baer, M. Golling, T. Südmeyer, U. Keller, Sesams for high-power oscillators: design guidelines and damage thresholds. IEEE J. Quantum Electron. **18**, 29–41 (2012)
21. E.R. Thoen, E.M. Koontz, M. Joschko, P. Langlois, T.R. Schibli, F.X. Kärtner, E.P. Ippen, L.A. Kolodziejski, Two-photon absorption in semiconductor saturable absorber mirrors. Appl. Phys. Let. **74**(26), 3927–3929 (1999)
22. R. Grange, M. Haiml, R. Paschotta, G. Spühler, L. Krainer, M. Golling, O. Ostinelli, U. Keller, New regime of inverse saturable absorption for self-stabilizing passively mode-locked lasers. Appl. Phys. B **80**, 151–158 (2005)
23. M. Larionov, *Kontaktierung und Charakterisierung von Kristallen für Scheibenlaser* (Herbert Utz Verlag, München, 2009)
24. J. Neuhaus, Passively mode-locked Yb:YAG thin-disk laser with active multipass geometry. PhD thesis, University Konstanz (2009)
25. V. Magni, Multielement stable resonators containing a variable lens. J. Opt. Soc. Am. A **4**(10), 1962–1969 (1987)
26. http://www.rp-photonics.com/powerscalingoflasers.html
27. V. Pervak, O. Pronin, O. Razskazovskaya, J. Brons, I.B. Angelov, M.K. Trubetskov, A.V. Tikhonravov, F. Krausz, High-dispersive mirrors for high power applications. Opt. Express **20**(4), 4503–4508 (2012)
28. V. Pervak, C. Teisset, A. Sugita, S. Naumov, F. Krausz, A. Apolonski, High-dispersive mirrors for femtosecond lasers. Opt. Express **16**(14), 10220–10233 (2008)
29. H. Stoehr, N. Rehbein, A. Douillet, J. Friebe, J. Keupp, T. Mehlstäubler, H. Wolff, E. Rasel, W. Ertmer, J. Gao, A. Giesen, Frequency-stabilized Nd:YVO4 thin-disk laser. Appl. Phys. B **91**(1), 29–33 (2008)
30. C. Teisset, Few-cycle high-repetition-rate optical parametric amplifiers and their synchronisation schemes. PhD thesis, TU Berlin.
31. H. Fattahi, C.Y. Teisset, O. Pronin, A. Sugita, R. Graf, V. Pervak, X. Gu, T. Metzger, Z. Major, F. Krausz, A. Apolonski, Pump-seed synchronization for MHz repetition rate, high-power optical parametric chirped pulse amplification. Opt. Express **20**(9), 9833–9840 (2012)
32. T. Udem, R. Holzwarth, T.W. Hänsch, Optical frequency metrology. Nature **416**, 233–237 (2002)
33. O. Pronin, J. Brons, C. Grasse, V. Pervak, G. Boehm, M.-C. Amann, V.L. Kalashnikov, A. Apolonski, F. Krausz, High-power 200 fs Kerr-lens mode-locked Yb:YAG thin-disk oscillator. Opt. Lett. **36**(24), 4746–4748 (2011)
34. V.L.Kalashnikov, *Solid State Lasers*, Chirped-pulse oscillators: Route to the energy-scalable femtosecond pulses pp. 145–184. InTech, (2012) http://www.intechopen.com/books/solid-state-laser/chirped-pulse-oscillators-a-route-to-the-energy-scalable-femtosecond-pulses
35. D.H. Sutter, G. Steinmeyer, L. Gallmann, N. Matuschek, F. Morier-Genoud, U. Keller, V. Scheuer, G. Angelow, T. Tschudi, Semiconductor saturable-absorber mirror assisted Kerr-lens mode-locked Ti:sapphire laser producing pulses in the two-cycle regime. Opt. Lett. **24**(9), 631–633 (1999)

Chapter 5
Kerr-Lens Mode-Locked Thin-Disk Oscillator

So far, all femtosecond thin-disk oscillators have been mode-locked by means of SESAM. The KLM technique has been proposed many times and simultaneously criticized as difficult to realize [1–3].[1] Actually, the limitations caused by SESAM[2] were the most motivating factor for experiments on the KLM TD oscillator. The main concerns and uncertainties in initiating this work were the natural coupling between SPM and self-focusing (the main mechanism for mode-locking), initiation of mode-locking and the influence of the TD thermal lens. As will be shown in this chapter, these concerns are not detrimental to successful mode-locking.

In Sect. 5.1 the basics of KLM are described, including self-focusing and state-of-the-art KLM resonators and experiments. Section 5.2 is devoted to the experimental work on Kerr-lens mode-locking of the TD Yb:YAG oscillator. The recipe to obtain KLM is presented here and oscillator characteristics are described. Section 5.3 shows oscillator stability in terms of output power, beam pointing and sensitivity to back reflections. Sections 5.4 and 5.5 briefly describe preliminary results on Kerr-lens mode-locking of Yb:YAG with positive GDD and Yb:Lu_2O_3 with negative GDD. The conclusions and further prospective developments are given in the last two sections.

5.1 Kerr-Lens Mode-Locking

5.1.1 KLM Basics

The refractive index n of a material depends on the incident electric field intensity I [4]

$$n = n_0 + n_2 I, \tag{5.1}$$

---

[1] This criticism is not truly objective and is sometimes of advertising and promotional character.

[2] At least, by the SESAMs available from Batop, ANU, Reflekron, WSI.

O. Pronin, *Towards a Compact Thin-Disk-Based Femtosecond XUV Source*,
Springer Theses, DOI: 10.1007/978-3-319-01511-8_5,

where n_2 is the nonlinear refractive index proportional to the third-order optical susceptibility $\chi^{(3)}$ [4]:

$$n_2 = \frac{16\pi^2\chi^{(3)}}{n_0 c} \tag{5.2}$$

and n_0 is the linear refractive index, c is the velocity of light. Formula 5.1 shows that light intensity modifies the refractive index, which, in turn, acts on the light, thus self-action of optical waves is possible: self-focusing, self-phase modulation, self-diffraction etc. In particular, self-focusing is important to understand the main principle of KLM. A Gaussian intensity distribution causes an increase of the refractive index in the central part of the beam in relation to its outer regions. The Gaussian beam then acquires a nonlinear phase shift of

$$\delta\varphi_{Kerr} = \frac{2\pi}{\lambda} n_2 I_0 \exp\left[-2\frac{r^2}{\omega^2}\right] l \approx \frac{2\pi}{\lambda} n_2 I_0 \left[1 - 2\frac{r^2}{\omega^2}\right] l,$$

where I_0 is the peak intensity on the axis, ω is the 1/e beam radius and l is the thickness of the Kerr medium. To first order, this expression can be taken as a parabolic function of $\frac{r^2}{\omega^2}$. By comparing it with the phase shift of a thin lens,

$$\delta\varphi_{Lens} = \frac{2\pi}{\lambda} n \frac{r^2}{f},$$

the expression for the focal length due to self-focusing yields

$$f = \frac{\omega^2}{2n_2 I_0 l} = \frac{\pi\omega^4}{4n_2 P l}, \tag{5.3}$$

where P is the peak power.

The higher the light intensity the stronger the action of such a lens. The lens becomes stronger for smaller beam radii ω and media with higher nonlinearity n_2. Self-focusing occurs for the high-intensity beam (red, Fig. 5.1) and reduces losses due to the hard-aperture blocking the low-intensity (CW) beam. This mechanism

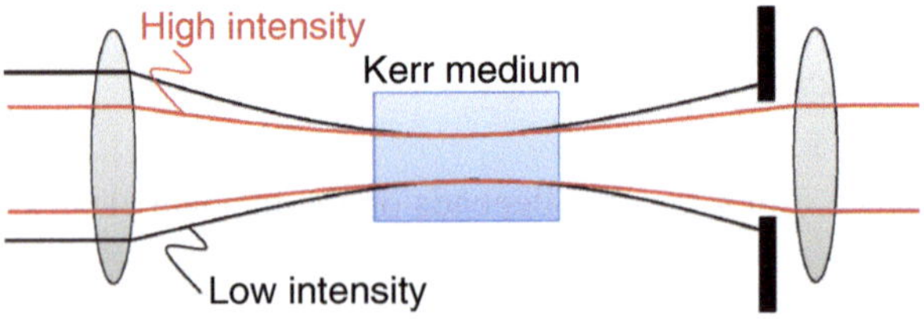

Fig. 5.1 Basic principle of KLM. Self-focusing occurs for a high-intensity beam (*red*) and reduces the losses due to a hard aperture (two *black knifes*) blocking the low-intensity (CW) beam. This mechanism initiates mode-locking and acts as an artificial saturable absorber

initiates mode-locking and acts as an artificial saturable absorber. Catastrophic run away damage can happen when a critical power P_{cr} is reached and the length of medium exceeds the self-focusing length z_{sf} according to [5, p.331]:

$$P_{cr} = \frac{\pi(0.61)^2\lambda^2}{8n_0n_2}, \qquad z_{sf} = \frac{2n_0\omega_0^2}{\lambda}\frac{1}{\sqrt{P/Pcr - 1}}. \tag{5.4}$$

The first oscillator working on the KLM principle was discovered by Spence et al. [6] and referred to as self-mode-locking or magic mode locking. Piché [7] explained the mode-locking mechanism on the basis of the self-focusing and only a few authors recognized the potential of self-focusing effect for mode-locking before the invention of KLM [8, 9]. Since that time KLM established itself as the method of choice for ultrashort-pulse generation and numerous studies [10, 11] were done on resonator design, theoretical numerical and analytical description of KLM and experiments on ultrashort pulse generation [12]. Mostly, experiments were performed with a Ti:Sa gain medium, which has several outstanding features: extremely broad bandwidth and short relaxations lifetime and high thermal conductivity [13].

5.1.2 KLM Resonator

The cavity configuration used for most mode-locking experiments is the bow-tie illustrated in Fig. 5.2 or so-called X-shape cavity or the Z-shape cavity. These types of resonators were analyzed in detail in [7, 14–16].

As one can see, the resonator configuration of the thin-disk oscillator in Fig. 4.9 is more complex in relation to that in Fig. 5.2. First, in Fig. 5.2 the Kerr medium simultaneously serves as a gain medium. This is not only hard to realize but also unnecessary in the TD configuration. It would strongly counteract the power-scaling concept, according to which the pump power density should be kept constant by enlarging the mode area on the disk and proportionally increasing the pump power. Moreover, the disk is too thin to exhibit reasonable self-focusing. There are a number of publications in which the gain medium and Kerr medium were separated. These are the experiments on Kerr-lens mode-locking of Nd:YAG [18] and Nd:YLF [19]

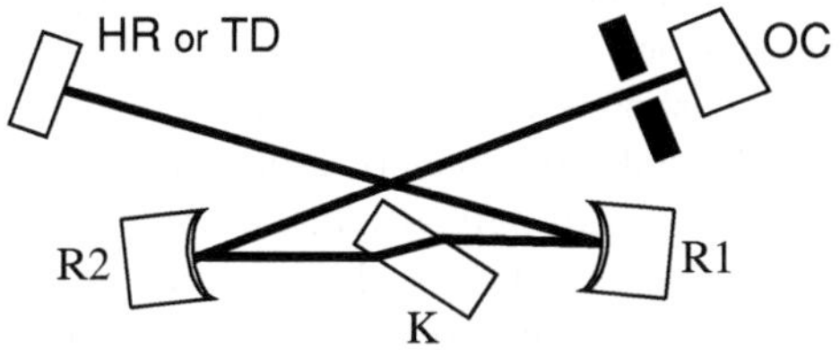

Fig. 5.2 Schematic of the X-shape resonator standardly used for low-power bulk oscillators. The replacement of the HR mirror by a TD leads to a design proposed in [17] for KLM realization in a thin-disk oscillator

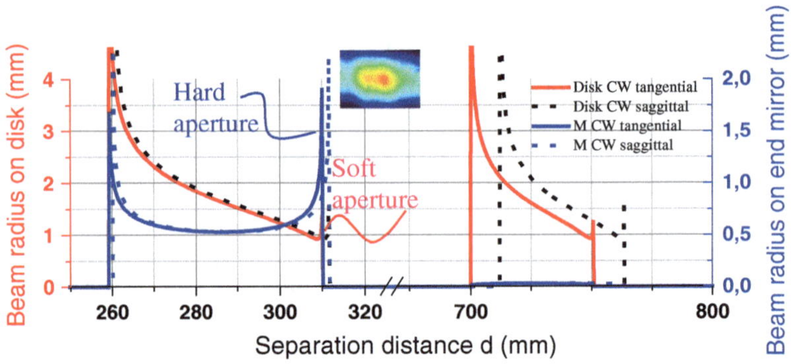

Fig. 5.3 Mode sizes on the disk and end mirror versus separation distance *d* for the CW cavity. The *inset* shows a strongly astigmatic beam profile at the stability edge corresponding to $d \sim 312$ mm

lasers. Indeed, they also have bow-tie-type cavities. This type of cavity is not used in the thin-disk configuration because of the small mode diameters over the optical elements, which is typically below 1 mm. The simple and convenient solution for KLM in the TD cavity was to implement a telescopic arrangement inside the cavity, which, in first approximation, cab be considered as a 4f extension. This resonator configuration is similar to that for a SESAM mode-locked case (see Figs. 4.9 and 5.3) and shows average mode diameters >2 mm over the intracavity optical elements.

5.1.3 Starting the KLM

One of the few drawbacks accompanying KLM is the difficult initiation of mode-locking. Small random initial intensity fluctuations are too weak to make the self-focusing effect pronounced. Thus, external cavity perturbation is necessary to induce stronger fluctuations and initiate pulse build-up. It can be accomplished by moving one of the cavity mirrors or placing one of the mirrors on a piezo transducer or shaker [20], or by using active mode-locking [19, 21], semiconductor-doped glass structures [22] or SESAM [23] for initial pulse build-up. In any case the oscillator has to operate near the stability edge. Exactly at the stability edge the oscillator is most susceptible to intensity fluctuations. This is typically accomplished by changing the distance between two focusing mirrors. In Fig. 5.4[3] the CW behaviour of the resonator at different stability edges is shown. Four stability edges appear as a function of the separation distance *d*. The stability edge corresponding to $d \sim 312$ mm was used in this work. The inset shows a strongly astigmatic CW beam profile at the stability edge, which agrees well with calculations. In principle, measurement of the beam profile at different cavity mirrors can be performed to derive the position with respect to the stability edge in the KLM regime.

[3] The calculations were done by Brons.

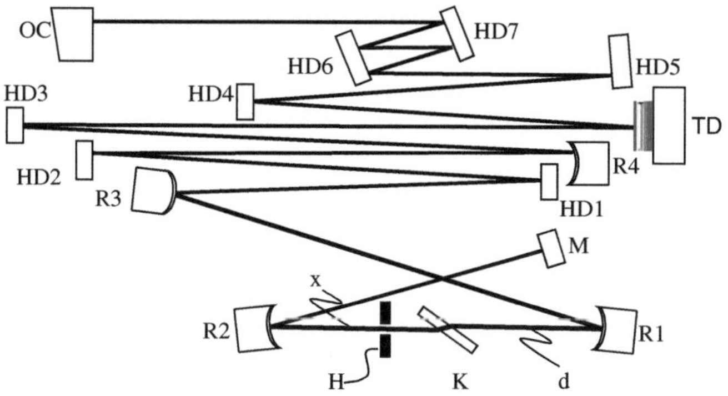

Fig. 5.4 Schematic of the KLM Yb:YAG disk oscillator in the presence of negative GDD. All flat mirrors are HD mirrors with $-1000\,fs^2$ and $-3000\,fs^2$ per bounce; OC, an output coupler with 5.5 or 14 % transmission; *R1*, *R2* concave mirrors with 0.3 m radius of curvature ; R3 $= -4$ m; R4 $= 4$ m; *K* Kerr medium; *H* Hard aperture; *M* Weak SESAM or HR mirror. The pulse repetition rate is 40 MHz

In some respect the starting problem is similar to the situation described in Sect. 4.4 for the SESAM mode-locked case when the mode-locker cannot be optimized for a whole range of energies and peak powers. Some weak mode-locking mechanism is needed to cover the low-energy range and pass on the baton to KLM for higher peak powers. Indeed, the situation concerning start-up is not very critical and is a subject for optimization. For example, self-starting stably mode-locked pure-KLM oscillators were demonstrated [14], and even turn-key commercial stable self-starting systems with $<$10 fs pulse duration are available [24]. It is worth noting that such an oscillator experiences a peak-power change of six to seven orders of magnitude and, moreover, self-starts.

5.1.3.1 Weak SESAM

SESAM used to be the intracavity element most susceptible to damage in high-power systems, typically via self-Q-switching or relaxation oscillation spikes during the pulse buildup process, as described in Sect. 4.4. Moreover, it also exhibits TPA, which tends to give rise to breakup into multi-pulsing and limits both the achievable intracavity pulse energy and minimum pulse duration. However, the use of a SESAM as starter for mode-locking strongly reduces constraints on the cavity design. In that case, it is no longer necessary to keep the saturable absorption large and relaxation time very short to maintain mode-locking. These considerations inspired us to design a SESAM more suitable to start mode-locking. We applied a dielectric top coating on a SESAM with a similar structure to that shown Fig. 4.1. The top reflective dielectric coating consists of six alternating layers of Ta_2O_5 (128 nm) and SiO_2 (178 nm). The coating reduces the intensity level in the SESAM by a factor of 9. As a consequence, the SESAM modulation depth drops by the same factor to $\Delta R < 0.1\,\%$ and the saturation fluence rises to $F_{sat} > 400\,\mu J/cm^2$, which leads to reduced TPA and an increased damage threshold [25]. Because of the small modulation depth, this SESAM alone cannot result in stable mode-locking. To our knowledge, such a low-modulation-depth weak SESAM was used as KLM starter for the first time [26].

In the experiments described below we used an OC with transmission of 5.5 % and a weak SESAM as end mirror M. The experiment with OC transmission of 14 % and HR as end mirror will be treated separately.

The analytical or numerical calculations of the Kerr-lens influence in our resonator can be very complex and, mostly, uncertain (or misleading) since exact values of some parameters such as the value of the TD thermal lens and astigmatism are missing. This together with very complex simulations of spatio-temporal dynamics of mode-locking could result in a huge uncertainties of these simulations. Therefore, as a first approximation we took the Kerr-lens as a thin lens equation (5.3) and used simple ABCD calculations to model the influence of the Kerr-lens. Such simulations gave a qualitative understanding of the cavity behaviour at the stability edge and possible influence of the Kerr lens on it. The first steps towards KLM were done experimentally.

5.2 Kerr-Lens Mode-Locking of Yb:YAG Oscillator

Kerr-lens mode locking of an Yb:YAG bulk oscillator was already performed by Uemura et al. and resulted in 35-fs pulses [27]. This remarkable performance came at the expense of dramatically reduced output power and efficiency, achieved by spectral filtering that shifted the central wavelength from 1030 nm toward 1060 nm (see Fig. 3.2), limiting the output power to 110 mW [26]. Even so, this can be compared with results for a similar bulk oscillator configuration mode-locked by a SESAM. In this case the shortest pulses were about 340 fs and the output power 110 mW [28].

The experimental setup of the KLM TD oscillator in Fig. 5.4 is similar to that shown in Fig. 4.9 for a SESAM mode-locked TD oscillator. The characteristic cavity dimensions and disk were kept the same. Dispersion from all intracavity elements as well as the nonlinear phase shift induced by the Kerr-medium are compensated for by the round-trip negative GDD of approximately $-22.000\,fs^2$. All flat mirrors were HD mirrors with -1000 and $-3000\,fs^2$ per bounce [29]. In order to provide high sensitivity of the Kerr effect, the oscillator is operated at the stability edge corresponding to the increased distance d between mirrors $R1$ and $R2$. This stability edge can be used for both soft- and hard-aperture KLM.

5.2.1 4-mm SF 57

In the beginning lead-containing SF 57 glass, with a high nonlinear refractive index $n_2 = 4.1 \cdot 10^{-19}\,m^2/W$ was chosen as Kerr medium to enhance the self-focusing effect and identify the resonator configuration and alignment necessary to obtain

KLM. Mirror R2 was mounted on a precision micrometer stage. A gradual increase of the distance d resulted in a gradual decrease of the output power and changed the beam profile to more astigmatic. Near the stability edge the output power substantially drops because of the poor overlap with the pump spot on the TD. This point serves as a reference near which Kerr lensing can be obtained. The mode-locking could be started by pushing mirror R2 and carefully aligning the position of the Kerr medium relatively to the focus (distance x in Fig. 5.4). The last alignment was not critical because a 4-mm-thick SF 57 plate can be placed in the focus with a small uncertainty. The distance d had to be increased to $\approx$312 mm in order to bring the cavity to the edge of the stability zone. It is clear that the beam radius of 50–100 μm in the Kerr medium and the high nonlinearity of SF 57 glass cause very high SPM, which cannot be balanced by the GDD of $-22000\,\mathrm{fs}^2$ introduced in the cavity. Unavoidably, this situation leads to unstable mode-locking with onset of multi-pulsing. Up to 10 pulses per cavity round-trip could be detected with a fast photodiode at 7 W output power and 70 W pump power (see Fig. 5.5a). A 5.5 % OC was chosen to decrease the lasing threshold and keep the intracavity power relatively large for initial experiments. Decreasing the pump power down to 50 and 3 W output power reduces the amount of

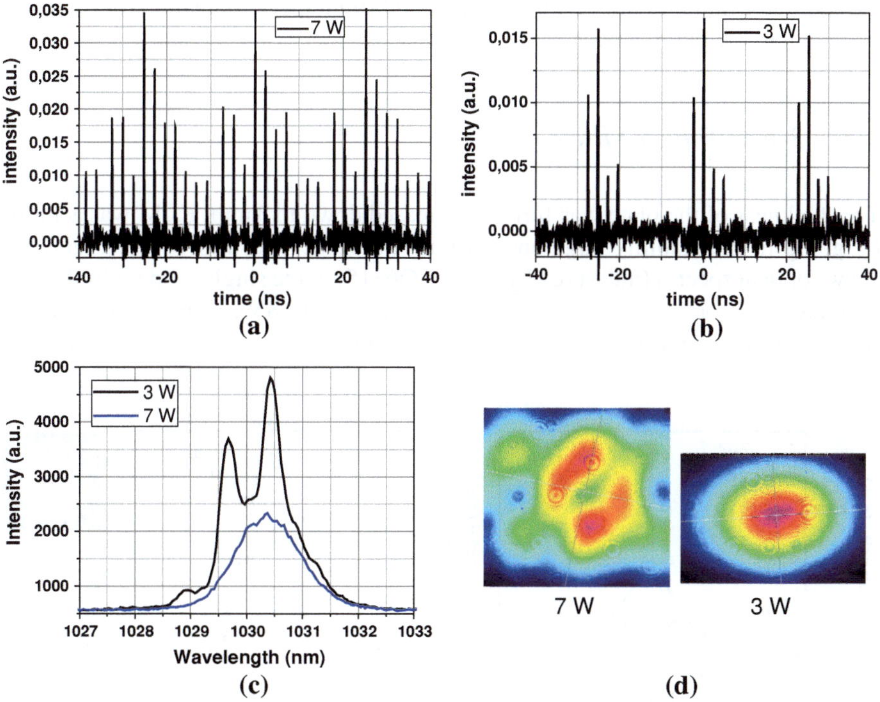

Fig. 5.5 Oscillator characteristics operating with 4-mm-thick SF 57 glass plate as Kerr medium. **a** Pulse train at 7 W output power; **b** Pulse train at 3 W output power; **c** Optical spectra at 3 and 7 W output power (measured with Ocean Optics HR 4000); **d** Beam profile at 3 and 7 W output power

pulses down to 3 (see Fig. 5.5b). The spectral bandwidth, however, does not change appreciably and maintains $\Delta\lambda_{FWHM} \approx 1.5$ nm (see Fig. 5.5c). Pronounced CW spikes appear at reduced output power. The thinner 1.8-mm plate of SF 57 glass was tried out as next step. It did not change the situation drastically as strong multi-pulsing and optical spectra with similar bandwidth were observed.

5.2.2 6.4-mm Fused Silica

Here, the 6.4-mm-thick fused silica plate with $n_2 = 2.5 \cdot 10^{-20}\,m^2/W$ [30] was taken. It reduced the nonlinearity by almost an order of magnitude in realtion to the 4-mm-thick SF 57 plate and simultaneously kept the plate thick in order to have no constraint on the alignment relatively to the beam focus. This could slightly improve the situation. In Fig. 5.6a the optical spectrum became broader with $\Delta\lambda_{FWHM} \approx 3$ nm for the mode-locked regime with onset of multiple pulses at 3 W and exhibited a pronounced CW component and narrower width $\Delta\lambda_{FWHM} \approx 2$ nm in the single-pulse regime. This regime is similar to the SESAM mode-locked case and slightly outperforms it in terms of pulse duration. Again, this was accomplished by starting the oscillator in the multi-pulsing regime and bringing it to the single-pulse regime by reducing the pump power.

5.2.3 3-mm Fused Silica

The thickness of the plate was further decreased down to 3 mm. KLM was achieved by going to the stability edge and pushing the stage with mirror *R2*. Figure 5.7a shows optical spectra (measured by Ando AQ6315) in the single- and multiple-pulse regimes with corresponding output powers of 4.5 and 8 W, respectively. Multiple pulses can be recognized by the strong modulation of the spectrum. The decrease

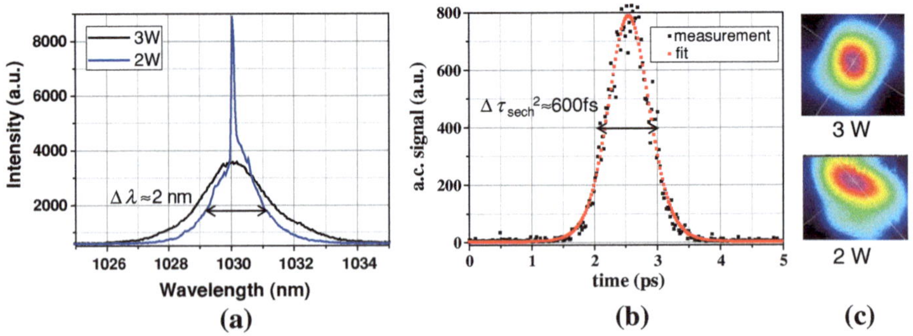

Fig. 5.6 Oscillator characteristics operating with a 6.4-mm-thick fused silica plate as Kerr medium. **a** Optical spectra at 3 and 2 W output power; **b** Autocorrelation trace measured at 2 W; **c** Beam profile at 3 and 2 W output power

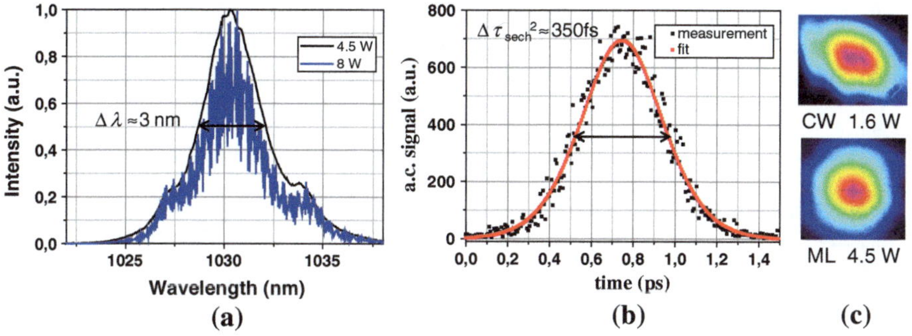

Fig. 5.7 Oscillator characteristics operating with 3-mm-thick fused silica plate as Kerr medium. **a** Optical spectra at 8 and 4.5 W output power (measured by ANDO AQ6315); **b** Autocorrelation trace measured at 4.5 W; **c** Beam profile in the CW and ML regimes. The noise in the *blue* trace is due to the long integration time

of the pump power leads to stable single-pulse operation at 4.5 W output power. There are two noticeable differences in comparison with previous regimes in the presence of strong nonlinearity. First, the spectral bandwidth $\Delta\lambda_{FWHM} \approx 3$ nm is much broader and accordingly pulses are shorter (Fig. 5.7b). Second, the difference in output power between the mode-locked and CW regimes becomes more pronounced (see Fig. 5.7c). This difference $P_{ML}/P_{CW} = 4.5\,\text{W}/1.5\,\text{W} = 3$ indicates a large effective modulation depth of the KLM mechanism.

5.2.4 1-mm Fused Silica

The best performance was achieved with a 1-mm-thick fused silica plate as Kerr medium. The oscillator was successfully operated with two different output couplers with 5.5 and 14 % transmission. With the lower output coupling, we achieved 17 W of average power at 110 W of pump power, which corresponds to an optical-to-optical efficiency of 15 %. Under these conditions, we measured a pulse duration of 200 fs (assuming a sech^2 pulse shape) with a spectral width of 6 nm (FWHM) and time-bandwidth product of 0.34 (ideal 0.315); see Fig. 5.8. Kelly sidebands are visible in the spectral wings. As can be seen the spectral bandwidth $\Delta\lambda_{FWHM} \approx 6$ nm approaches the emission-bandwidth limit of Yb:YAG (see Figs. 3.2 and 5.8) and the difference in output power between the mode-locked and CW regimes is now very pronounced (see Fig. 5.8c). This difference $P_{ML}/P_{CW} = 17\,\text{W}/2\,\text{W} = 8$ indicates the large effective modulation depth of the KLM mechanism, which according to the simulations[4] is $\Delta R_{KLM} = 5\,\%$. With high output coupling, we could achieve stable operation up to pump power levels of 180 W, resulting in an average output power of 45 W, corresponding to 1.1-μJ pulses and 25 % optical-to-optical efficiency. The autocorrelation and spectral shapes are shown in Fig. 5.9 with a pulse duration

[4] Preliminary simulations were done by Kalashnikov.

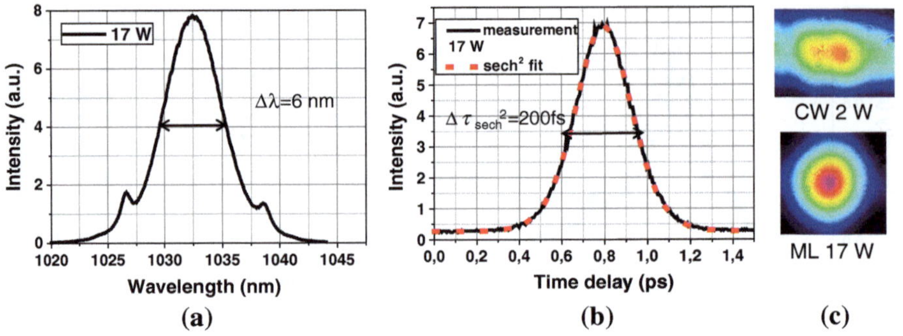

Fig. 5.8 Oscillator characteristics operating with 1-mm-thick fused silica plate as Kerr medium. **a** Optical spectrums at 17 W; **b** Autocorrelation trace at 17 W; **c** Beam profile in the CW and ML regimes

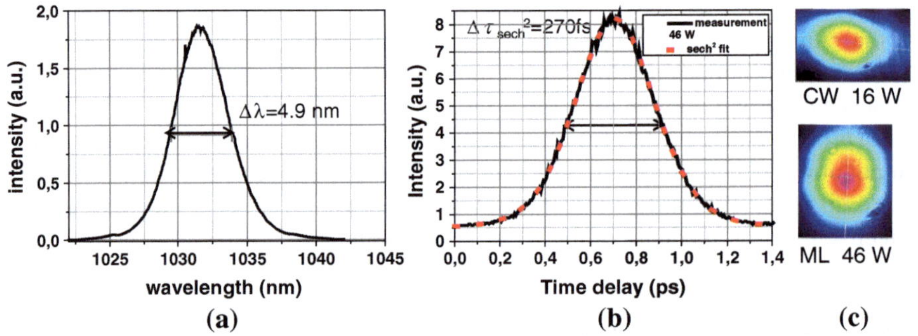

Fig. 5.9 Oscillator characteristics operating with 1-mm-thick fused silica plate and 14 % OC. **a** Optical spectra at 46 W; **b** Autocorrelation trace at 46 W; **c** Beam profile in CW and ML regime

of 270 fs, a spectral width of 4.9 nm (FWHM) and the time-bandwidth product of 0.37 (ideal 0.315) [26]. In both operation regimes, the intracavity pulse energy was approximately 8 μJ and the intracavity average power was 320 W. The attempts to reach higher energy led to the onset of multipulsing or CW background. Again, the strong hysteresis behaviour of the oscillator was utilized to start and reach a stable operation regime. At high pump power levels of approximately 150 W for 5.5 % OC and 220 W for 14 % OC the oscillator self-started in multiple pulses or with CW background; afterwards, the pump power was decreased to bring the oscillator into stable single-pulse operation.

Q-switched mode-locking instabilities used to appear only during the alignment of the oscillator. The suppression of these instabilities is very likely caused by saturation of the KLM self-amplitude modulation (see Fig. 2.4).

A small red shift of the spectrum relatively to the gain maximum at 1030 nm could be observed. This shift amounts to 1–2.5 nm and is more pronounced for higher intracavity peak powers. It can be explained by the action of radiation reabsorption in combination with gain saturation and dispersion of the gain medium [31, 32]. Typically, the optimal cavity alignment and mode-locking regime corresponded to a maximal red shift of the spectrum.

It is worth noting that the cavity in the CW regime is not sensitive to the alignment of such a thin plate as long as its position x relatively to the focus is fixed. The light refraction in the thin plate is weak, thus small misalignment from the Brewster angle due to a plate exchange or due to a shift along the focus is not critical for the overall cavity alignment.

5.2.5 1-mm Fused Silica, Pure KLM, 5.5 % OC

The above-described experiments were carried out with the weak SESAM as a starter for mode-locking. In order to check the influence of a weak SESAM on the oscillator performance, it was replaced (see position M in Fig. 5.4) by an HR mirror. The first experiment was carried out without a hard aperture inside the resonator and KLM was realized due to the soft-aperture on the TD pump spot. This oscillator configuration showed high sensitivity to the position d at the stability edge as well as difficult start-up. Critical sensitivity to the alignment and difficult start-up were decisive factors for not using this configuration for further experiments. An additional hard-aperture in the form of a pinhole was added into the cavity between the focus and concave mirror *R2* (see Fig. 5.4). The hard-aperture drastically reduced the alignment sensitivity of the oscillator and simplified the starting procedure. The oscillator could easily be started by pushing mirror *R2* or self-started at increased pump power levels, as described for weak-SESAM-assisted KLM. In the first experiments a simple 1-mm-thick copper plate with a drilled hole >2 mm in diameter in it was used as a hard-aperture. Due to the high intracavity power >300 W it quickly warmed up and caused an oscillator drift. In order to eliminate this issue a water-cooled monolithic copper aperture (Fig. 5.10c) was recently designed and used in later experiments. This water-cooled aperture does not cause pronounced drift of the oscillator, but it also hardly

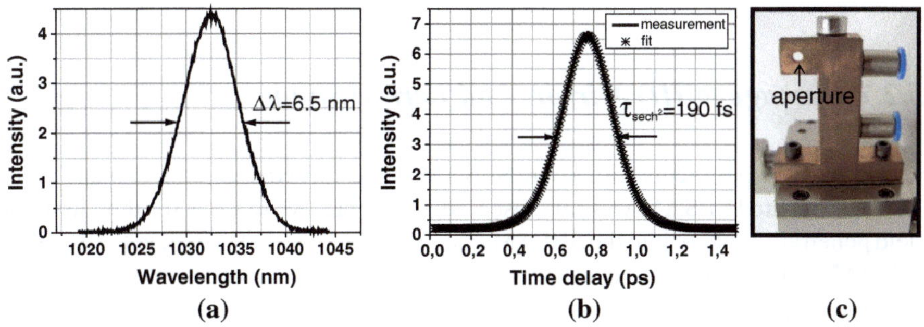

Fig. 5.10 Oscillator characteristics operating with 1-mm-thick fused silica plate, 5.5 % OC and hard-aperture. **a** Optical spectrums at 15 W; **b** Autocorrelation trace at 15 W; **c** Water-cooled hard-aperture

tolerates such oscillator drifts due to the TD thermal lens. Thus, several alignments may be necessary due to the oscillator warm-up. Once aligned, the oscillator keeps running stably for the whole day. With the low output coupling of 5.5 % we achieved 15 W of average power. Under these conditions, we measured a pulse duration of 190 fs (assuming a $sech^2$ pulse shape) with a spectral width of 6.5 nm (FWHM) in Fig. 5.10a, which is 10 % broader than that reported for the SESAM-assisted case. This pulse shortening by 10 % is due to the combined action of the soft and hard apertures with somewhat higher effective modulation depth. This argument can also be checked by the absence of Kelly sidebands in the optical spectrum in Fig. 5.10a.

5.2.6 0.5-mm Fused Silica

Here, a test experiment was made by exchanging the 1-mm-thick plate for a 0.5-mm-thick plate. The alignment relatively to the beam focus became more critical. The oscillator was not self-starting as described before, but slight mirror perturbation was sufficient to initiate KLM. The regime showed single-pulse operation with characteristics similar to the case with the 1-mm plate. The main outcome of this experiment was to demonstrate operation with a reduced SPM threshold and higher output power. Obviously, further experiments on decreasing the thickness of the Kerr medium and/or increasing the beam size in it should be continued in order to reach the maximum output energy at a minimum threshold SPM value.

It is important to note that all experiments described in this work were performed with the disk TD 157 (Table 3.1). It could be run near the stability edge at pump powers >250 W and experience the sudden change of the intracavity power due to the action of the Kerr lens (see Fig. 5.8), showing robust performance. Damage only occurred when the pump power was unintentionally increased very rapidly.[5] The exchange for a new disk TD 159 (Table 3.1) was done on the same day with no noticeable difference in oscillator performance. This proves that the cavity is insensitive to the initial variation in radius of curvature of the TD, as can be seen in the stability diagrams in Fig. 3.5.

5.2.7 Damage to HD Mirrors and Kerr Medium

Unlike in the experiments with a pure SESAM mode-locked oscillator, damage of the weak overcoated SESAM was not observed. There are two reasons for this: first, the field penetration-depth into the SESAM structure is by a factor 9 less than for the usual SESAM with no dielectric coating; second, the beam on the end mirror is enlarged near the stability edge and consequently reduces the peak intensity on it. However,

[5] The control box had maximum control voltage at the moment of turning on the pump diodes. The maximum control voltage corresponds to a pump power of >500 W.

Table 5.1 Table summarizing the advantages and disadvantages of KLM and SESAM-ML

SESAM	KLM
−Dependent on wavelength	+Independent of wavelength
−Real absorber with saturable and non-saturable losses	+Artificial absorber, Kerr medium does not warm up
−TPA	+ TPA-free
−Finite relaxation time (<1 ps)	+Ultrashort relaxation time <10 fs
−Moderate modulation depth for high power operation (<2 %)	+Modulation depth (>5 %)
−Degradation	+No degradation
−Low damage threshold	+High damage threshold
−Difficult growth, characterization, poor reproducibility	+No growth, no characterization
−Expensive	+Cheap
+Slight constraints on the cavity design	−Serious constraints on the cavity design
+Often self-starting	−Often not self-starting
+Low nonlinearity	−High nonlinearity due to SPM in the Kerr medium

damage to the HD mirrors and the Kerr medium was observed. This damage was only observed during oscillator start-up and alignment. It was mostly caused both by the non-optimal position x of the Kerr medium relatively to the focus and by moving the oscillator too close to the stability edge via changing the distance d. It was noticed that HD mirrors located closer to the output coupler and that those with higher GDD values were most susceptible to damage. The beam diameter shrinks near the stability edge in the part of the oscillator close to the output coupler and makes the optics in this part most susceptible to damage during the oscillator start-up. As long as the oscillator is optimally aligned it can be started and run with no damage to the optical elements.

5.2.8 Energy Scaling of KLM

KLM as realized in our work allows energy/power scaling. The average power scaling is supported by the TD concept and the cavity calculations show that spot sizes over 5 mm on the TD are feasible. Energy scaling can be performed by increasing the mode size inside the Kerr medium in proportion to the increase of the intracavity energy. This procedure would keep the peak intensity and self-focusing parameters in the Kerr medium constant. This can be accomplished in turn by exchanging the mirrors *R1* and *R2* for new ones with larger radii of curvature. The mirrors *R1*, *R2* separated by the distance d can be considered as a telescopic 4f arrangement in first approximation. Consequently, exchanging the mirrors will not influence the beam parameters on other critical cavity elements such as the TD. Further optimization of KLM can be accomplished by changing the thickness of the Kerr medium.

5.3 Oscillator Stability

5.3.1 Beam Pointing

The oscillator is operating in the stability zone II [33] with large mode sizes. Despite this, the laser shows excellent beam pointing stability with $<10\,\mu$rad r.m.s. deviation, as can be derived from the measurements in Fig. 5.11b. Figure 5.11a shows the beam pointing drift during oscillator warm-up, which takes 10–30 min. Beam pointing fluctuations correspond to the level of commercial Ti:Sa oscillators, and so no active stabilization schemes are needed for further experiments with this laser.

5.3.2 Intensity Fluctuations

Many research applications such as CEP stabilization [34], spectrum broadening [35], white light generation, repetition rate synchronization [36] and seeding of enhancement cavities [37] call for unprecedented power stability of the oscillator, specifically on a short time scale. All these experiments utilize nonlinear phenomena, thus any small intensity fluctuations of a driving field can be transferred to the amplitude and

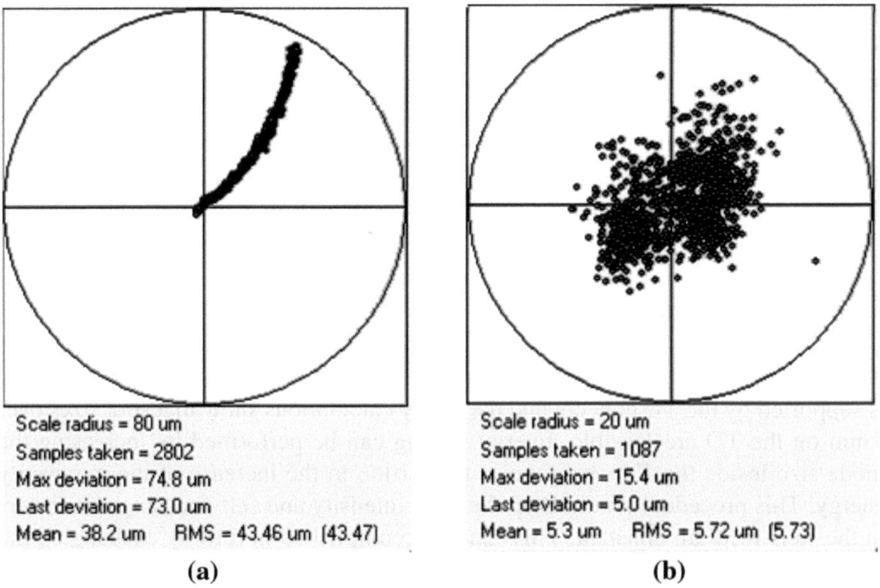

Fig. 5.11 Beam pointing fluctuations of the mode-locked oscillator at 1 m distance from the output coupler (measured with WinCam beam profiler). **a** Beam pointing drift during oscillator warm-up; **b** Beam pointing fluctuations after oscillator warm-up

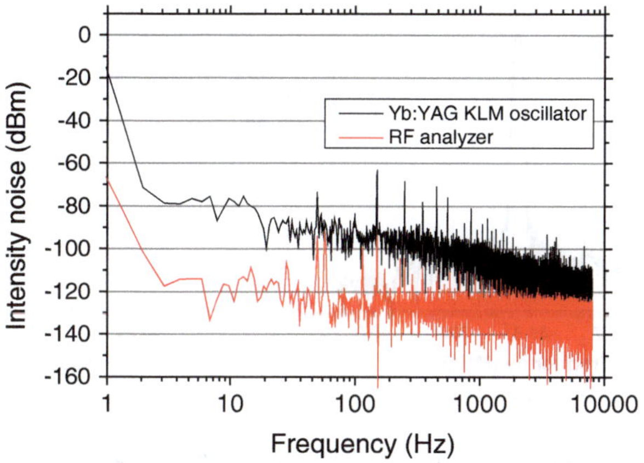

Fig. 5.12 One-sided intensity power spectral density (PSD) of intensity noise from the free-running Yb:YAG TD KLM oscillator (*black curve*) and the noise of the Agilent spectrum analyzer (*red curve*)

phase noise of the output signal. The oscillator intensity noise is typically caused by fluctuations of the pump diodes and by external mechanical noise. The power spectral density of the intensity fluctuations is shown in Fig. 5.12. The main intensity noise lies within the 10 kHz bandwidth and has a small relaxation oscillation spike near 8 kHz. Corresponding peak-to-peak intensity fluctuations are below 0.8 % and are defined by the fluctuations of the pump diodes. This result approaches the noise level of commercial Ti:Sa oscillators. It is difficult to give an adequate comparison between the intensity noise of the SESAM mode-locked TD oscillator and the KLM TD oscillator as the first one was mode-locked at 3 times higher intracavity pulse energies then the former, but with similar intracavity peak powers of 40 MW. Generally the SESAM mode-locked oscillator had a stronger tendency to Q-switched mode-locking, which was not observed in the KLM oscillator. The simplest way of characterizing intensity fluctuations was to detect the repetition rate frequency with an RF spectrum analyzer. As shown in Fig. 5.13, the sidebands in the case of the SESAM mode-locked oscillator are suppressed by 60 dBm and sidebands of the KLM oscillator are suppressed by 70–75 dBm. It is worth noting that suppression of the sidebands by 40 dBm was reported for a SESAM-ML active multi-pass oscillator [38, 39]. This shows a remarkable superiority of KLM over SESAM mode-locking.

5.3.3 Sensitivity to Back Reflections

Typically, back reflections disturb the oscillator operation and can cause strong fluctuations or interrupt mode-locking. Measures should be taken to avoid back reflections from all intracavity elements, and so TD and OC should be wedged. Moreover, a Faraday isolator may be necessary to prevent back reflections from the experimental

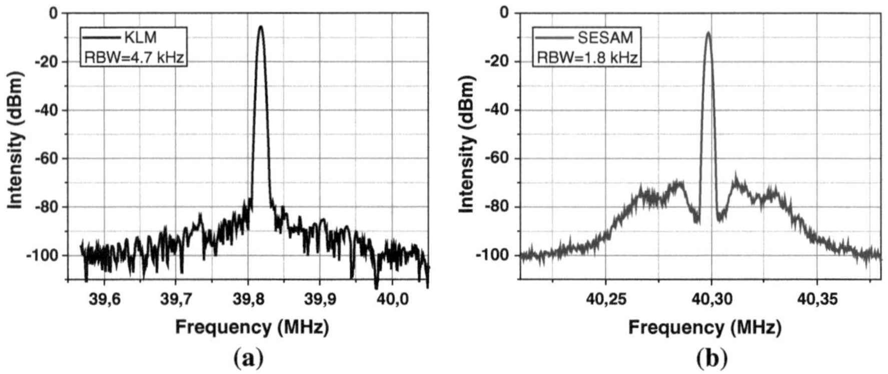

Fig. 5.13 The radio frequency spectra showing the repetition rate beat signal at ~40 MHz. **a** KLM Yb:YAG TD oscillator; **b** SESAM mode-locked Yb:YAG TD oscillator

setup outside the cavity. The sensitivity of the oscillator to back reflections was tested by placing a plane-parallel output coupler with 2 % transmission having an AR coating <0.1 % instead of the wedged one. The oscillator could easily be mode-locked at 6 W output power by pushing the translation stage with mirror *R2* (see Fig.5.4); no pulse train modulation was noticed. Moreover, the mode-locking could not be disturbed even by placing a HR mirror after the output coupler and reflecting back all of the 6 W output power. Further experiments with spectrum broadening in a fibre (see Sect. 6.1.1) and CE phase characterization were performed without the Faraday isolator and with no influence on the oscillator performance. The influence of back reflections on the self-starting oscillator behaviour [40] was not studied.

5.3.4 Mode Quality at the Stability Edge

One of the concerns with KLM oscillators can be the mode quality. As demonstrated in Fig. 5.8, the astigmatic shape of the mode near the stability edge in CW operation changes to a TEM_{00} fundamental one in the mode-locked regime. This strongly depends on the oscillator position at the stability edge (separation distance *d* in Fig. 5.4). A first contribution to mode quality deterioration can arise from the fact that self-focusing does not push it far enough away from the stability edge, where the sagittal and tangential planes do not overlap perfectly, so that the mode may still have an astigmatic shape. The second reason can be a frequency degeneracy of higher transverse modes near the stability edge [41–43]. Modes with an M^2 parameter from 1.1 to 1.6 were observed, depending on how close to the stability edge the oscillator was operating. The frequency degeneracy near the stability edge combined with mirror damage (as a spatial filter) may result in strong coupling of higher-order transverse modes. Figure 5.14 demonstrates the beam profile in the CW regime near the cavity stability edge when one of the mirrors was damaged. Such spatial filters as

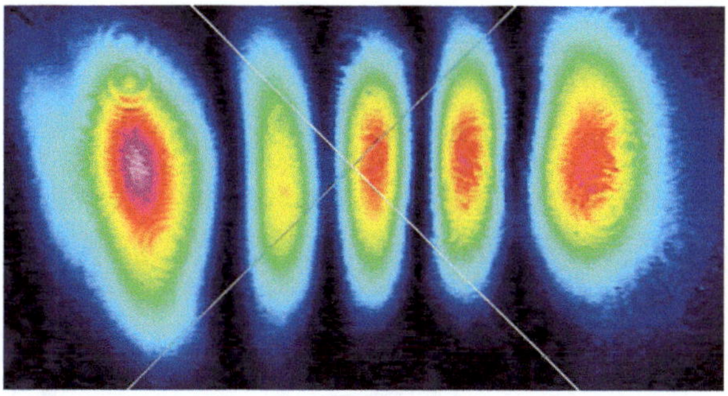

Fig. 5.14 TEM_{04}-like mode profile measured in CW regime for the cavity near the stability edge when one of the mirrors was damaged

damaged spots or dust should be avoided in order to suppress the coupling of higher transverse modes. The self-starting or easy-to-start oscillator regime indicates a safe position at the stability edge which does not cause reasonable beam deterioration.

Some applications, may benefit much, for example, from the radially polarized donut-like mode [44]. XUV generation inside the oscillator can be organized in a similar manner to that proposed in [41] and transverse mode degeneracy can also be used as an XUV outcoupling method.

5.4 Kerr-Lens Mode-Locked Yb:YAG Thin-Disk Chirped-Pulse Oscillator

The regime of positive intracavity GDD has already demonstrated its potential for energy scaling of long-cavity Ti:Sa oscillators [45], reaching pulse energies of >500 nJ [46] and more recently >650 nJ in commercial systems [24]. This approach has also been applied to SESAM mode-locked Yb:KYW and Yb:KLuW bulk [47] and thin-disk [48] oscillators allowing a substantial increase of the intracavity pulse energy. Nevertheless, while constituting the standard operation regime of femtosecond fibre oscillators [49], positive GDD is still an exception to the rule in bulk solid-state oscillators. As a follow-up of our work on the Yb:YAG KLM TD oscillator in the negative GDD regime, here the operation of such a system is described in the regime of positive intracavity GDD [50]. The experimental setup, sketched in Fig. 5.15, is practically the same as the one in Fig. 5.4 for the case of the negative GDD regime. At first, pure KLM was tried by using a hard-aperture. However, all attempts to start this mode of operation proved unsuccessful [as opposed to the system operated with negative GDD, where hard-aperture KLM could be readily started (Sect. 5.2)]. Therefore, the results reported here were achieved with soft-aperture KLM and with the assistance of a weak SESAM employed merely for self-starting mode-locking [50].

As discussed in Sect. 2.3, positive intracavity GDD has several advantages over the negative one. The peak power inside the oscillator is drastically reduced owing to

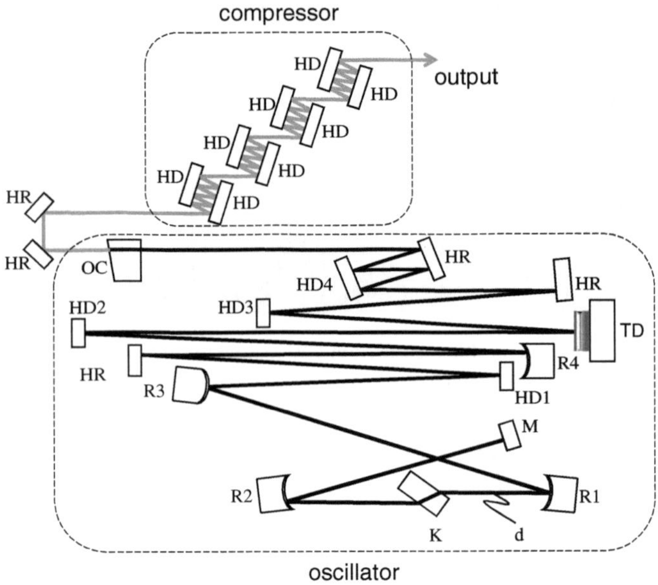

Fig. 5.15 Schematic of the KLM mode-locked Yb:YAG disk oscillator in the presence of positive GDD. All flat mirrors are HD mirrors with $+250$ and $+1000\,\mathrm{fs}^2$ per bounce; OC, an output coupler with 5.5 or 10 % transmission; *R1, R2* concave mirrors with 0.3 m radius of curvature ; R3, -4 m; R4, 4 m; BP, 6.4-or 9.5-mm-thick fused silica Brewster plate; M, weak SESAM. The pulse repetition rate is 40 MHz

large pulse chirp and remains constant with the energy scaling. As a result, the GDD needed to stabilize a pulse is approximately an order of magnitude less than needed in negative GDD regime. This is expected to allow shorter pulses (after extracavity compression) because the Fourier limit is predicted to scale with the square root of the intracavity GDD [51]. Alternatively, at the same magnitude of GDD, the positive operation regime is expected to allow higher maximum pulse energy than the negative GDD regime, since the energy scaling in the presence of positive GDD is superior to that with negative GDD [51].

These benefits come at the expense of a more challenging start-up behaviour of the mode-locking process. In practice, pulse formation with positive GDD can only be initiated within a limited parameter window (of the SPM and GDD values). In contrast to negative-GDD KLM, which tends to start in the multiple-pulse regime, positive intracavity GDD tends to prefer the formation of a single pulse in the cavity. As compared with our KLM disk laser operated originally with a net intracavity GDD of $-22000\,\mathrm{fs}^2$ in Sect. 5.2, a positive GDD nearly an order of magnitude smaller was chosen while leaving other parameters of the cavity unchanged. The round-trip GDD of $+3500\,\mathrm{fs}^2$ was introduced by 4 bounces on dispersive mirrors. This modification of the cavity did not allow mode-locking. Therefore, subsequent increase of SPM inside the cavity was performed by replacing the 1-mm-thick fused-silica nonlinear medium with one having a thickness of a quarter of an inch (6.4 mm).

5.4.1 Oscillator Start-up

The pulse build-up process is rather hard and starts most likely from Q-switching. This Q-switching often resulted in white light generation and occasionally damaged the Kerr medium but never damaged any other optics in the cavity. The transit situation, when the oscillator is close to the optimal point and tries to build up a pulse train, is shown in Fig. 5.16. A small pump power increase leads to a stable pulse train, depicted by the red curve in Fig. 5.16a. The peak intensity is at least six times higher during the oscillator start-up than for the stable situation (compare black and red traces). The optical spectrum corresponding to the start-up situation is shown in Fig. 5.16b. It exhibits both a CW component and small broad pedestal corresponding to the mode-locked pulse train. The experiments were carried out with output couplers of 5.5 and 10 % transmission and 6.4- and 9.5-mm-thick fused-silica Kerr media. Generally, the tendency for softer self-start and alignment flexibility were observed for the case of the 5.5 % OC and 6.4-mm-thick Kerr plate. It is worth noting that strong Q-switched mode-locking was also observed several years ago and prevented us from mode-locking a thin-disk oscillator in the regime of positive GDD with a SESAM only because of damage to the optical components.

5.4.2 Results

In this configuration, with an output coupling of 5.5 %, 17 W of average power at 150 W of pump power was achieved, which corresponds to an optical-to-optical efficiency of 11 %. Under these conditions, a pulse duration of 1.7 ps (assuming a $sech^2$ pulse shape) with a full spectral width of over 20 nm (see red spectrum in Fig. 5.17) was measured. The calculated Fourier limit is 150 fs. The pulses were compressed

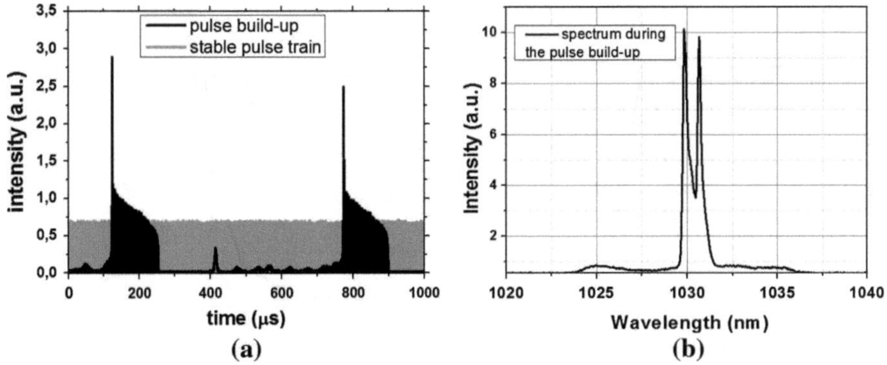

Fig. 5.16 Oscillator self-starting behaviour. **a** Pulse train during build-up (*black*) and mode-locked pulse train (*red*); **b** Spectrum corresponding to the pulse build-up

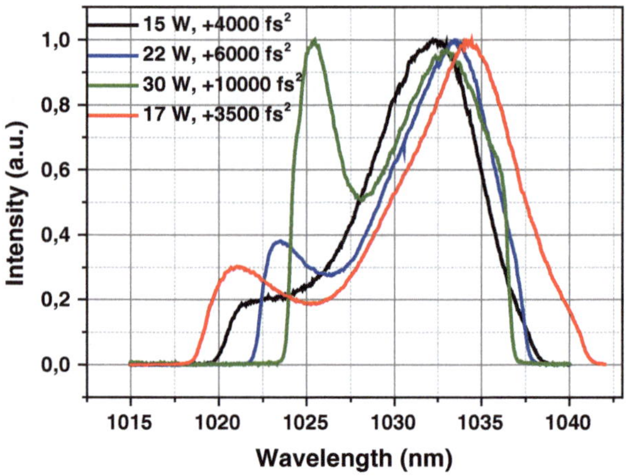

Fig. 5.17 The spectra at different output power levels. The red spectrum was measured at 17 W of output power, a 5.5 % output coupler and a 6.4-mm-thick Kerr plate. The other spectra were measured with 10 % output coupler and 9.5-mm-thick Kerr plate

externally in an all-dispersive mirror compressor (Fig. 5.15) introducing $-60.000\,\text{fs}^2$ GDD at 20 bounces [29]. The resulting pulse duration after the compressor is 190 fs full-width at half-maximum (assuming a sech^2 pulse profile) with 11 W throughput (Fig. 5.18). The relatively high loss of 40 % in the compressor is caused by clipping the beam on the mirrors and is related to the limited amount of mirrors [50]. The pedestal in the autocorrelation trace (Fig. 5.18) may be caused by the high-dispersive mirrors, which were designed for the range 1030 ± 10 nm and could be slightly shifted in the central wavelength because of coating deposition errors [29]. Depending on the mode-locking parameters, the spectrum can exhibit strong non-linear chirp at the edges, a common feature of the positive dispersion regime [50]. The autocor-

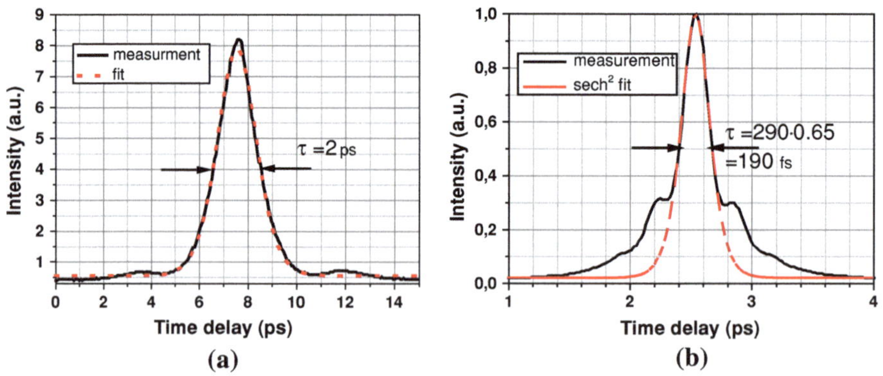

Fig. 5.18 The autocorrelation measurements before compressor (**a**) and after compressor (**b**)

relation with pedestal may be due to an uncompensated non-linear chirp [52]. The spectral shapes in Fig. 5.17 have relatively smooth edges as a result of the narrow gain bandwidth and the relatively large net loss, (see [46, 51]) and become more truncated with the growth of the GDD. Also, there is an asymmetry of the spectra relatively to the emission bandwidth maximum at ~1030 nm. This can be explained by a similarly asymmetric shape of the Yb:YAG gain (Sect. 3.2) and phase change in such a narrow-band gain medium [53].

After the oscillator self-started it was slightly realigned to operate closer to the stability edge by means of increasing the distance d (see Fig. 5.15) and proportionally the pump power. This optimization procedure typically provided an increase of the average power by 30 %; see blue, olive and black spectra in Fig. 5.17, measured with 10 % output coupler transmission and 9.5-mm-thick Kerr plate. An extreme case is the red spectrum in Fig. 5.17, measured with 5.5 % output coupling and a 6.4-mm-thick Kerr plate. In this case the output power could be increased by 50 %, from 12 to 17 W. A slight increase of the pump power by 3–5 % with all other parameters fixed resulted in a tendency to instabilities and strong Q-switching, similar to that in the build-up process (see Fig.5.16a). Decrease of the pump power by the same amount always brought the laser back into the stable regime (see Fig. 5.16b). This was characterized by typical peak-to-peak amplitude fluctuations of <3 % within the 1 min measurement time, a value significantly higher than that for the negative GDD regime [50] (see Sect. 5.3.2).

As was predicted in [51], a positive GDD regime in combination with soft-aperture KLM can allow powerful energy scaling without the need for a substantial GDD increase and comprising bandwidth. Nevertheless, we found experimentally that the energy growth increasingly impedes the ability of the laser to self-start and requires an increase of the GDD level for self-starting KLM. It is this GDD growth required for self-starting that in fact results in spectral narrowing (Fig. 5.17). This behaviour is similar to the one reported in [54]. The oscillator could substantially benefit from the ability to tune the GDD inside the cavity and use this 'knob' to implement a hysteresis in the *energy* versus *GDD* coordinates. This was realized by means of two prisms [55]. Our attempts to initiate mode-locking at an increased GDD level $>+10000\,\text{fs}^2$ were not successful. Further power scaling of the oscillator with positive GDD was limited by the inability to start mode-locking [50].

5.4.3 Conclusion

Even though the difficult mode-locking initiation can be explained qualitatively by the long chirped pulses, being especially critical for KLM, the details and theoretical investigation of the starting mechanism remains one of the most intriguing and non-investigated issues for energy-scalable oscillators. This work provides strong motivation for theoretical research on the starting mechanism, specifically in the presence of positive GDD. 0.4 μJ pulses with an average power of 17 W and a pulse duration of 190 fs were demonstrated. This result is realized with nearly an order smaller GDD compared with the regime of negative GDD (in Sect. 5.2). Power scaling with GDD and output coupling up to 30 W demonstrated, however, approaching GDD levels similar to the negative dispersion regime (Fig. 5.17). Inability to start the oscillator at increased GDD levels was found to be the limiting factor in further energy scaling. An additional starting mechanism such as regenerative mode-locking [21] or SESAM [23] with an increased modulation depth offers the potential for scaling the pulse energy and the power of KLM disk lasers to unprecedented values [50].

5.5 Kerr-Lens Mode-Locking of Yb:Lu_2O_3 Oscillator

The generation of near emission-bandwidth-limited pulses from Yb:YAG prompted experiments on Kerr-lens mode-locking of the more-broadband gain medium Yb:Lu_2O_3. The designed resonator was of convex-concave type similar to the resonator described in Sect. 5.2 with an identical focusing arrangement consisting of $R1 = R2 = 300$ mm concave mirrors and having similar misalignment sensitivity. The pump head was aligned for 24 passes, providing a pump spot 2.8 mm in diameter on the Yb:Lu_2O_3 disk. The initial (unpumped) disk curvature was $\sim$2 m. We used VBG-stabilized pump diodes lasing at 976 nm with a FWHM emission bandwidth of $\sim$2 nm for efficient pumping (see Fig. 3.2).

5.5.1 Technical Obstacles

Four disks were available for these experiments (from the group of Huber). All of them were soldered to metal heat sinks. Two disks with a large diameter of 10 mm were from the same boule and two smaller-diameter disks were from a different boule, grown at a somewhat earlier time.

5.5.1.1 Yb:Lu_2O_3 Disks with 10-mm Diameter

Preliminary diagnostic[6] of the 10-mm disks in a microscope with cross-polarizers showed crystal inhomogeneities for both disks. In Fig. 5.19 both Yb:Lu_2O_3 disks are shown for comparison. One of the disks was edge-bevelled after soldering. This resulted in small damage of the disk near the edge. This disk was destroyed under moderate pumping of below 50 W.

The second large disk was tested in a multi-mode CW cavity and showed no lasing. Just replacing the output coupler by an HR mirror could allow lasing in the multimode regime, which at least indicates high losses introduced by the disk. Moreover, partial darkening of the pump spot was observed. In particular, at increased pump

[6] The pictures were made by Larionov.

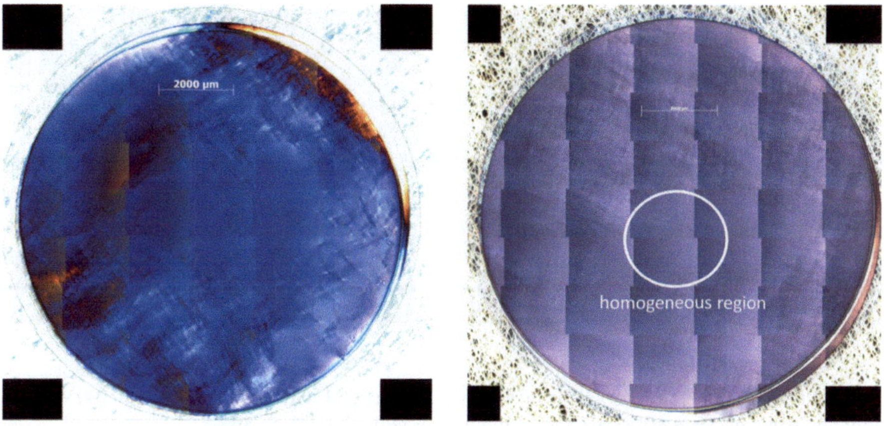

(a) (b)

Fig. 5.19 Cross-polarized microscope images of two 10-mm-diameter $Yb{:}Lu_2O_3$ disks

powers (>100 W) part of the pump spot became dark, which completely distorted the laser performance. The disk could withstand up to 200 W pump power, exhibiting this darkening effect. Better performance could not be reached by changing the pump spot position on the disk.

5.5.1.2 $Yb{:}Lu_2O_3$ Disks with 4-mm Diameter

The disks 4 mm in diameter showed substantially better performance in comparison with the larger disks. Thus, in the multi-mode regime the optical-to-optical efficiency of 50 % could easily be reached while pumping with 200 W. Nevertheless, another technical problem appeared: the pump spot diameter on the disk could not be reduced below 2.8 mm. This is due to the relatively large 1-mm-diameter coupling fibre of the pump diodes and constraints on the size of the collimating optics. Moreover, the disks had parts of the solder at the edges, which absorbed fluorescence light and heated up the disk. The temperature rise over 50 °C between the pump spot and the disk edges is plainly visible in Fig. 5.20. Operation of both disks under these conditions is risky, because it may lead to disk damage and induces strong thermal disk distortions, which degrade the oscillator performance in the fundamental mode regime.

5.5.2 Results

A 1-mm-thick glass plate was chosen as Kerr medium and introduced a round-trip dispersion of $-10000\,\mathrm{fs}^2$. By gradually increasing the distance d between mirrors

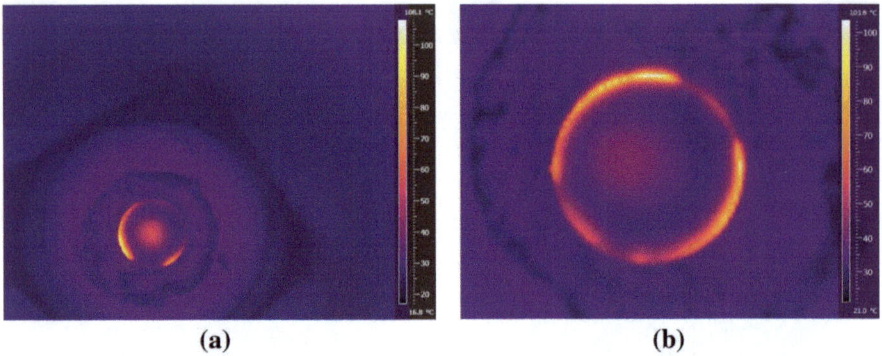

Fig. 5.20 The temperature gradient of 4-mm-diameter Yb:Lu_2O_3 disks under the 100 W pump with no lasing. **a** 1 % at. disk; **b** Zoomed pump spot region of 3 % at. disk

R1 and *R2*, the resonator was brought to the stability edge. A hard-aperture was installed between mirror *R2* and the focus. No SESAM was used for this experiment in order to completely eliminate uncertainties from its curvature and thermal lens. By pushing mirror *R1* mounted to the translation stage the oscillator could be routinely mode-locked. The oscillator was running with rather poor mode quality and strong beam pointing fluctuations were induced by the strong thermal aberrations in the disk. Neither the CW nor the ML regime showed much difference in terms of these fluctuations. In this mode of operation an output power of 7.5 W could be achieved with 3.5 % OC and ~130 W pump power. The pulses are rather long with ~430 fs duration (Fig. 5.21).

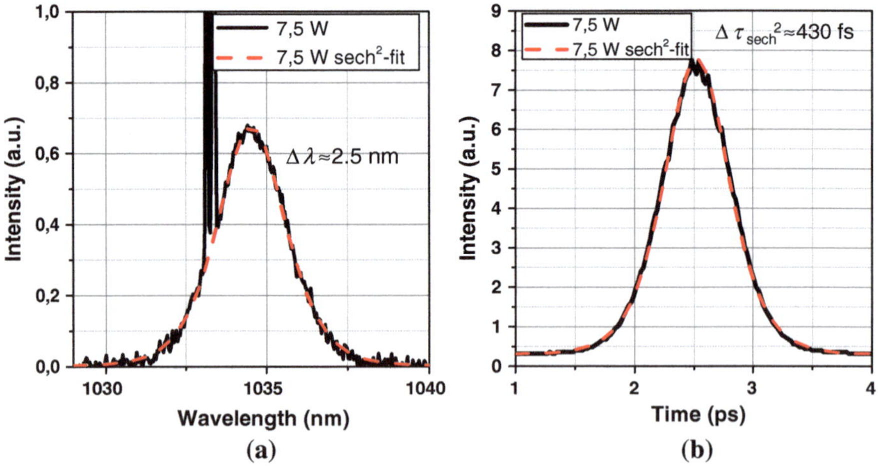

Fig. 5.21 Yb:Lu_2O_3 oscillator characteristics operating with 1-mm-thick fused-silica plate, 3.5 % OC and hard-aperture. **a** Optical spectrum at 7.5 W with CW component; **b** Autocorrelation trace at 7.5 W

5.5.3 *Conclusion*

The preliminary results on the mode-locking of the Yb:Lu_2O_3 oscillator are rather moderate. Nevertheless, KLM could be realized even with the strong thermal problems from the disk and consequently poor mode quality and large beam pointing fluctuations. Historically, this is the second thin-disk oscillator, mode-locked via a Kerr lens. Decreasing the pump spot size on the disk or working with larger disk diameters will dramatically improve the situation. The reproducibility of the KLM approach and robustness is main outcome of this experiment.

5.6 Comparison of KLM and SESAM-ML Techniques

The advantages and disadvantages of both the SESAM and KLM techniques are summarized here. The objectivity of this summary is supported by the use of both techniques in the identical Yb:YAG oscillator setup.

5.7 Further Steps

5.7.1 *Pulse Shortening*

Ways of shortening the pulse duration from the thin-disk KLM oscillator can be accomplished conventionally by using a broader Yb-doped gain medium such as Yb:Lu_2O_3, Yb:KYW, Yb:$LuScO_3$ Yb:CALGO, and/or by increasing the modulation depth of the KLM mechanism and thus generating pulses beyond the emission bandwidth limit. This scenario is fully supported by the theoretical description. One could consider the realization of a Ti:Sa thin-disk geometry as it becomes possible with the availability of compact high-average-power green pump sources [56].

Slight suppression of the gain maximum can be applied to flatten the gain spectrum for the purpose of reaching an effectively broader gain bandwidth. Proper multilayer coating design of an HR mirror may provide such a discrimination[7] around the wavelength corresponding to the gain maximum. This suppression should not be as strong as in [27] in order to work close to the gain maximum.

The effective gain bandwidth is always truncated by losses in the resonator (see Sect. 2.2.7). Reduction of these losses, mainly due to the output coupler, helps to exploit the whole emission bandwidth of the gain medium. On the other hand, it unavoidably reduces the output power and increases the intracavity peak power. However, this is even beneficial to intracavity-based nonlinear experiments such as XUV generation [57].

[7] Private discussion with Pervak.

5.7.2 Energy Scaling

Cryogenic cooling can be applied to increase the gain and reduce thermal effects (Sect. 3.3.2) in the TD [58]. This approach has the strong drawback of reducing the emission bandwidth [59] and thus increasing the minimal achievable pulse duration.

Zero-phonon-line pumping can be used to reduce the quantum defect. The progress in the manufacturing of volume Bragg gratings [60] makes high-power pump sources (>800 W) commercially available [61]. For instance, the Yb:YAG zero-phonon-line is centred around 969 nm and has $\Delta\lambda_{FWHM} \approx 2$ nm. Pumping at this wavelength reduces the quantum defect by 30 %[8] in comparison with a pumping wavelength of 940 nm. Thus, roughly, 30 % less heat is deposited in the Yb:YAG gain medium.

The active multi-pass cavity [39] in combination with KLM can be a good approach for further energy scaling. First, it substantially increases the gain and allows high output coupling rates to be used, thus decreasing the intracavity peak power and consequently constrains on all optical elements. Second, the intrinsic advantages of the KLM mechanism (see Table 5.1) support energy scaling.

Solving the problems with the start-up of the positive dispersion regime opens a route to unprecedentedly high energies.

The access to several-100-μJ energies in sub-200-fs pulses directly from the oscillator is feasible with a combination of the above-listed conditions in the positive dispersion regime.

References

1. U. Keller, Recent developments in compact ultrafast lasers. Nature **424**, 831–838 (2003)
2. R. Paschotta, U. Keller, Ever higher power from mode-locked lasers. Opt. Photon. News **14**(5), 50–54 (2003)
3. U. Keller, Ultrafast solid-state laser oscillators: a success story for the last 20 years with no end in sight. Appl. Phys. B **100**, 15–28 (2010)
4. S.A. Akhmanov, SYu. Nikitin, *Physical Optics* (Clarendon Press, Oxford, 1997)
5. R.W. Boyd, *Nonlinear Optics*, 3rd edn. (Academic, Boston, 2007)
6. D.E. Spence, P.N. Kean, W. Sibbett, 60-fsec pulse generation from a self-mode-locked Ti:sapphire laser. Opt. Lett. **16**(1), 42–44 (1991)
7. M. Piché, Beam reshaping and self-mode-locking in nonlinear laser resonators. Opt. Commun. **86**(2), 156–160 (1991)
8. E.G. Lariontsev, V.N. Serkin, Possibility of using self-focusing for increasing contrast and narrowing of ultrashort light pulses. Sov. J. Quantum Electron. **5**(7), 796 (1975)
9. M. Marconi, O. Martinez, F. Diodati, Short pulse generation in solid state lasers by a novel passive technique. Opt. Commun. **63**(3), 211–216 (1987)

8

$$\Delta q = \frac{969\text{–}940\,\text{nm}}{1030\text{–}940\,\text{nm}} = 0.32.$$

10. F. Krausz, M. Fermann, T. Brabec, P. Curley, M. Hofer, M. Ober, C. Spielmann, E. Wintner, A. Schmidt, Femtosecond solid-state lasers. IEEE J. Quantum Electron. **28**(10), 2097–2122 (1992)
11. T. Brabec, F. Krausz, Intense few-cycle laser fields: frontiers of nonlinear optics. Rev. Mod. Phys. **72**, 545–591 (2000)
12. F. Krausz, M. Fermann, T. Brabec, P. Curley, M. Hofer, M. Ober, C. Spielmann, E. Wintner, A. Schmidt, Beam reshaping and self-mode-locking in nonlinear laser resonators. IEEE J. Quantum. Electron. **28**(10), 2097–2122 (1992)
13. P.F. Moulton, Spectroscopic and laser characteristics of Ti:Al_2O_3. J. Opt. Soc. Am. B **3**(1), 125–133 (1986)
14. G. Cerullo, S.D. Silvestri, V. Magni, Self-starting Kerr-lens mode locking of a Ti:sapphire laser. Opt. Lett. **19**(14), 1040–1042 (1994)
15. D. Huang, M. Ulman, L.H. Acioli, H.A. Haus, J.G. Fujimoto, Self-focusing-induced saturable loss for laser mode locking. Opt. Lett. **17**(7), 511–513 (1992)
16. V.L. Kalashnikov, V.P. Kalosha, I.G. Poloyko, V.P. Mikhailov, Optimal resonators for self-mode locking of continuous-wave solid-state lasers. J. Opt. Soc. Am. B **14**(4), 964–969 (1997)
17. E. Wintner, E. Sorokin, I. Sorokina, Laser system for producing ultra-short light pulses. US Patent 6363090B1 (2002)
18. B. Henrich, R. Beigang, Self-starting Kerr-lens mode locking of a Nd:YAG-laser. Opt. Commun. **135**, 300–304 (1997)
19. G.P.A. Malcolm, A.I. Ferguson, Self-mode locking of a diode-pumped Nd:YLF laser. Opt. Lett. **16**(24), 1967–1969 (1991)
20. Y.M. Liu, K.W. Sun, P.R. Prucnal, S.A. Lyon, Simple method to start and maintain self-mode-locking of a Ti:sapphire laser. Opt. Lett. **17**(17), 1219–1221 (1992)
21. L. Turi, F. Krausz, Amplitude modulation mode locking of lasers by regenerative feedback. Appl. Phys. Lett. **58**(8), 810–812 (1991)
22. I.P. Bilinsky, R.P. Prasankumar, J.G. Fujimoto, Self-starting mode locking and Kerr-lens mode locking of a Ti:Al_2O_3 laser by use of semiconductor-doped glass structures. J. Opt. Soc. Am. B **16**, 546–549 (1999)
23. D.H. Sutter, G. Steinmeyer, L. Gallmann, N. Matuschek, F. Morier-Genoud, U. Keller, V. Scheuer, G. Angelow, T. Tschudi, Semiconductor saturable-absorber mirror assisted Kerr-lens mode-locked Ti:sapphire laser producing pulses in the two-cycle regime. Opt. Lett. **24**(9), 631–633 (1999)
24. http://www.femtolasers.com/Oscillator.50.0.html
25. C.J. Saraceno, C. Schriber, M. Mangold, M. Hoffmann, O.H. Heckl, C.R.E. Baer, M. Golling, T. Südmeyer, U. Keller, Sesams for high-power oscillators: design guidelines and damage thresholds. IEEE J. Quantum Electron. **18**, 29–41 (2012)
26. O. Pronin, J. Brons, C. Grasse, V. Pervak, G. Boehm, M.-C. Amann, V.L. Kalashnikov, A. Apolonski, F. Krausz, High-power 200 fs Kerr-lens mode-locked Yb:YAG thin-disk oscillator. Opt. Lett. **36**(24), 4746–4748 (2011)
27. S. Uemura, K. Torizuka, Sub-40-fs pulses from a diode-pumped Kerr-lens mode-locked Yb-doped yttrium aluminum garnet laser. Jap. J. Appl. Phys. **50**(1), 010201 (2011)
28. C. Hönninger, R. Paschotta, M. Graf, F. Morier-Genoud, G. Zhang, M. Moser, S. Biswal, J. Nees, A. Braun, G. Mourou, I. Johannsen, A. Giesen, W. Seeber, U. Keller, Ultrafast ytterbium-doped bulk lasers and laser amplifiers. Appl. Phys. B **69**, 3–17 (1999)
29. V. Pervak, O. Pronin, O. Razskazovskaya, J. Brons, I.B. Angelov, M.K. Trubetskov, A.V. Tikhonravov, F. Krausz, High-dispersive mirrors for high power applications. Opt. Express **20**(4), 4503–4508 (2012)
30. R. Adair, L.L. Chase, S.A. Payne, Nonlinear refractive index of optical crystals. Phys. Rev. B **39**, 3337–3350 (1989)
31. V.L. Kalashnikov, E. Sorokin, I.T. Sorokina, Mechanisms of spectral shift in ultrashort-pulse laser oscillators. J. Opt. Soc. Am. B **18**(11), 1732–1741 (2001)
32. V.L. Kalashnikov, E. Sorokin, S. Naumov, I.T. Sorokina, Spectral properties of the Kerr-lens mode-locked Cr^{4+}: YAG laser. J. Opt. Soc. Am. B **20**(10), 2084–2092 (2003)

33. V. Magni, Multielement stable resonators containing a variable lens. J. Opt. Soc. Am. A **4**(10), 1962–1969 (1987)
34. S.A. Meyer, J.A. Squier, S.A. Diddams, Diode-pumped Yb:KYW femtosecond laser frequency comb with stabilized carrier-envelope offset frequency. Eur. Phys. J. D **48**(1), 19–26 (2008)
35. T. Ganz, V. Pervak, A. Apolonski, P. Baum, 16 fs, 350 nJ pulses at 5 MHz repetition rate delivered by chirped pulse compression in fibers. Opt. Lett. **36**(7), 1107–1109 (2011)
36. H. Fattahi, C.Y. Teisset, O. Pronin, A. Sugita, R. Graf, V. Pervak, X. Gu, T. Metzger, Z. Major, F. Krausz, A. Apolonski, Pump-seed synchronization for MHz repetition rate, high-power optical parametric chirped pulse amplification. Opt. Express **20**(9), 9833–9840 (2012)
37. A. Cingöz, D.C. Yost, T.K. Allison, A. Ruehl, M.E. Fermann, I. Hartl, J. Ye, Direct frequency comb spectroscopy in the extreme ultraviolet. Nature **482**, 68–71 (2012)
38. D. Bauer, I. Zawischa, D.H. Sutter, A. Killi, T. Dekorsy, Mode-locked Yb:YAG thin-disk oscillator with 41 μJ pulse energy at 145 W average infrared power and high power frequency conversion. Opt. Express **20**, 9698 (2012)
39. J. Neuhaus, D. Bauer, J. Zhang, A. Killi, J. Kleinbauer, M. Kumkar, S. Weiler, M. Guina, D.H. Sutter, T. Dekorsy, Subpicosecond thin-disk laser oscillator with pulse energies of up to 25.9 microjoules by use of an active multipass geometry. Opt. Express **16**(25), 20530–20539 (2008)
40. K. Tamura, J. Jacobson, E.P. Ippen, H.A. Haus, J.G. Fujimoto, Unidirectional ring resonators for self-starting passively mode-locked lasers. Opt. Lett. **18**(3), 220–222 (1993)
41. J. Weitenberg, P. Rußbüldt, T. Eidam, I. Pupeza, Transverse mode tailoring in a quasi-imaging high-finesse femtosecond enhancement cavity. Opt. Express **19**(10), 9551–9561 (2011)
42. R. Paschotta, Beam quality deterioration of lasers caused by intracavity beam distortions. Opt. Express **14**(13), 6069–6074 (2006)
43. Q. Zhang, B. Ozygus, H. Weber, Degeneration effects in laser cavities. Eur. Phys. J. Appl. Phys. **6**, 293–298 (1999)
44. M.A. Ahmed, M. Haefner, M. Vogel, C. Pruss, A. Voss, W. Osten, T. Graf, High-power radially polarized Yb:YAG thin-disk laser with high efficiency. Opt. Express **19**(6), 5093–5103 (2011)
45. S.H. Cho, F.X. Kärtner, U. Morgner, E.P. Ippen, J.G. Fujimoto, J. Cunningham, W.H. Knox, Generation of 90-nJ pulses with a 4-MHz repetition-rate Kerr-lens mode-locked Ti:Al_2O_3 laser operating with net positive and negative intracavity dispersion. Opt. Lett. **26**(8), 560–562 (2001)
46. S. Naumov, A. Fernandez, R. Graf, P. Dombi, F. Krausz, A. Apolonski, Approaching the microjoule frontier with femtosecond laser oscillators. New J. Phys. **1**, 216 (2005)
47. G. Palmer, M. Emons, M. Siegel, A. Steinmann, M. Schultze, M. Lederer, U. Morgner, Passively mode-locked and cavity-dumped Yb:$KY(WO_4)_2$ oscillator with positive dispersion. Opt. Express **15**(24), 16017–16021 (2007)
48. G. Palmer, M. Schultze, M. Siegel, M. Emons, U. Bünting, U. Morgner, Passively mode-locked Yb:$KLu(WO_4)_2$ thin-disk oscillator operated in the positive and negative dispersion regime. Opt. Lett. **33**(14), 1608–1610 (2008)
49. F. Wise, A. Chong, W. Renninger, High-energy femtosecond fiber lasers based on pulse propagation at normal dispersion. Laser Photon Rev. **2**(1–2), 58–73 (2008)
50. O. Pronin, J. Brons, C. Grasse, V. Pervak, G. Boehm, M.-C. Amann, A. Apolonski, V.L. Kalashnikov, F. Krausz, High-power Kerr-lens mode-locked Yb:YAG thin-disk oscillator in the positive dispersion regime. Opt. Lett. **37**(17), 3543–3545 (2012)
51. V.L. Kalashnikov, Chirped-pulse oscillators: route to the energy-scalable femtosecond pulses, in *Solid State Lasers*, ed. by A.H. Al-Khursan (InTech, 2012), pp. 145–184
52. V. Kalashnikov, E. Podivilov, A. Chernykh, S. Naumov, A. Fernandez, R. Graf, A. Apolonski, Approaching the microjoule frontier with femtosecond laser oscillators. New J. Phys. **1**, 217 (2005)
53. R. Paschotta, R. Häring, A. Garnache, S. Hoogland, A. Tropper, U. Keller, Soliton-like pulse-shaping mechanism in passively mode-locked surface-emitting semiconductor lasers. Appl. Phys. B **75**, 445–451 (2002)

54. A. Fernandez, Chirped-pulse oscillators: generating microjoule femtosecond pulses at megahertz repetition rate. Ph.D. thesis, LMU München (2007)
55. A. Fernandez, T. Fuji, A. Poppe, A. Fürbach, F. Krausz, A. Apolonski, Chirped-pulse oscillators: a route to high-power femtosecond pulses without external amplification. Opt. Lett. **29**(12), 1366–1368 (2004)
56. http://www.ipgphotonics.com/Green_CW_Laser.htm
57. E. Seres, J. Seres, C. Spielmann, Extreme ultraviolet light source based on intracavity high harmonic generation in a mode locked Ti:sapphire oscillator with 9.4 MHz repetition rate. Opt. Express **20**(6), 6185–6190 (2012)
58. N. Vretenar, T.C. Newell, T. Carson, P. Peterson, T. Lucas, W.P. Latham, H. Bostanci, J.J. Huddle-Lindauer, B.A. Saarloos, D. Rini, Cryogenic ceramic 277 watt Yb:YAG thin-disk laser. Opt. Eng. **51**(1), 014201 (2012)
59. D. Brown, R. Cone, Y. Sun, R. Equall, Yb:YAG absorption at ambient and cryogenic temperatures. IEEE J. Quantum Electron. **11**(3), 604–612 (2005)
60. B.L. Volodin, S.V. Dolgy, E.D. Melnik, E. Downs, J. Shaw, V.S. Ban, Wavelength stabilization and spectrum narrowing of high-power multimode laser diodes and arrays by use of volume Bragg gratings. Opt. Lett. **29**(16), 1891–1893 (2004)
61. http://www.dilas.de

Chapter 6
Towards Ultrashort CE Phase Stable Pulses

As discussed in Sect. 3.2, the most promising gain media for thin-disk operation are Yb:YAG and $Yb:Lu_2O_3$ with the corresponding gain bandwidths of $\Delta\lambda_{FWHM,YAG}$ $\approx$9 nm and $\Delta\lambda_{FWHM,Lu_2O_3}$ $\approx$13 nm. Even the realization of gain-bandwidth-limited operation of an $Yb:Lu_2O_3$ oscillator hardly accesses the sub-100-fs regime. Methods of external pulse shortening are inevitably needed in order to reach shorter pulses. Such methods have existed for a long time now and are based on, for instance, the action of SPM on the pulse spectrum during propagation in a medium with a certain Kerr-nonlinearity.

The generation of short highly energetic pulses from a TD-based laser setup allows many applications [1]. The range of research application drastically increases when these pulses can be shortened to several cycles and be CEP-stabilized.

This chapter[1] is structured as follows: the basics of spectral broadening in a fibre, state-of-the-art experiments on it and results with the KLM TD oscillator are given in Sect. 6.1. Then, in Sect. 6.2 the setup for CE phase measurements and a discussion of the arising difficulties are presented. Finally, origins of CE phase noise, preliminary results on stabilization via control of the diode current and further steps are described in Sect. 6.4.

6.1 Spectral Broadening in Fibre

In Sect. 2.2, devoted to soliton mode-locking, the influence of SPM on the spectrum was described. The nonlinear Schrödinger equation (2.13) governs the propagation of pulses inside a medium with some nonlinearity γ [$W^{-1}km^{-1}$], losses and group velocity dispersion (GVD) β_2 [ps^2/km]. Depending on β_2 and γ, a pulse can evolve differently, and so it is convenient to introduce two parameters: dispersion length L_D and nonlinear length L_{NL} [2]:

[1] The results described in this chapter were obtained in close collaboration with Fabian Lücking.

O. Pronin, *Towards a Compact Thin-Disk-Based Femtosecond XUV Source*,
Springer Theses, DOI: 10.1007/978-3-319-01511-8_6,

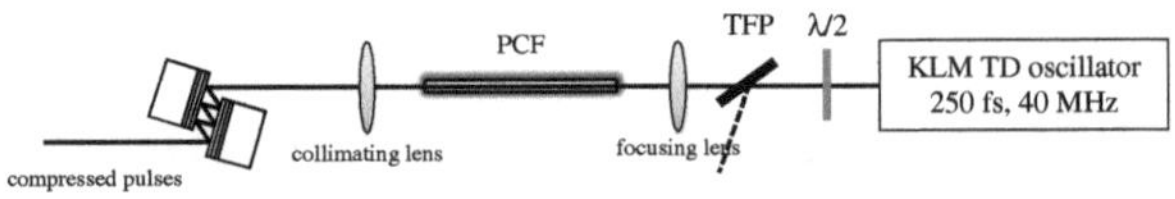

Fig. 6.1 Schematic of the setup for spectral broadening in PCF. $\lambda/2$ half-wave plate; *TFP* thin-film polarizer; *PCF* photonic crystal fibre

$$L_D = \frac{\tau^2}{|\beta_2|}, \qquad L_{NL} = \frac{1}{\gamma P_0}. \tag{6.1}$$

These parameters can be compared with the propagation length L to show over what length scale dispersive and nonlinear effects are important. The optimal propagation wavelength z_{opt} and compression ratio $t_{in}/t_{compressed}$ can also be defined via these two parameters [3].

SPM-induced spectral broadening requires high intensities over a long pulse propagation distance. This can be satisfied by propagating the light through a fibre for which Photonic Crystal Fibres (PCF) [4, 5] are excellent candidates. They provide large core sizes exceeding a diameter of 35 μm and simultaneously support a single transverse mode and a large spectral range. To date, many experiments on spectral broadening in PCFs have been performed with high-repetition-rate laser systems [6–10]. The PCFs utilized in these works are typically under 10 cm in length and have core diameters <35 μm, while the pulse energies stay below 1 μJ. Among those experiments with Yb-laser systems the Yb-fibre CPA system used in work [8] is closest to our oscillator. It delivers 270-fs pulses with 65 W average power at 78 MHz repetition rate. Spectral broadening of this output was performed in a PCF, 35 μm in diameter and 6 cm in length and resulted in a 57-W throughput with 27-fs pulses. This work thus gives at least a hint about the upper limit in pulse duration that can be reached with the oscillator.

6.1.1 One-Stage Compression at 40 MHz

The Yb:YAG KLM oscillator, described in Sect. 5.2, is used and provides on a daily basis 45 W of output power and 250-fs pulses at 40 MHz repetition rate. A half-wave plate and thin-film polarizer are placed directly after the oscillator to allow continuous power attenuation (see Fig. 6.1). The beam is focused into the fibre via a lens with focal distance f = 30 mm and collimated by an achromatic lens (AC254, Thorlabs) with f = 30 mm. The large-mode-area (LMA) PCF, LMA 25 and LMA 35 with 25-μm and 35-μm core diameter (NKT Photonics), were used for spectral broadening. The fibres are placed in a V-groove aluminium plate, fixed to a precise 3D translation stage. It is worth noting that no Faraday isolator was placed in between the oscillator and fibre. First, the maximal achievable spectral broadening as well as maximal power which can be coupled into the 6-cm-long LMA 25 fibre, were

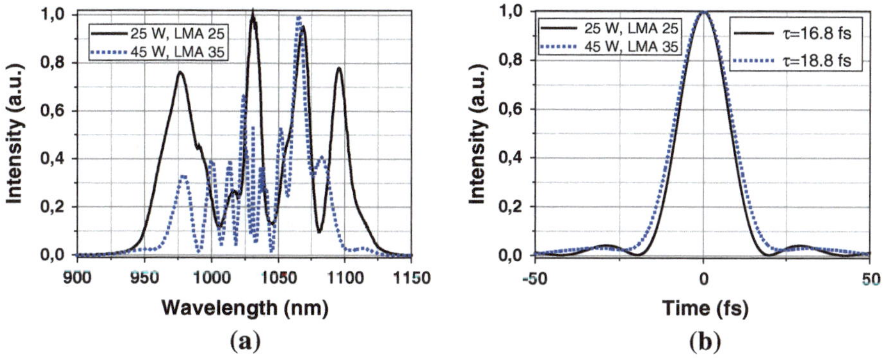

Fig. 6.2 Spectral broadening achieved with 25 W and 45 W input power for LMA 25 and LMA 35 fibres (**a**) and corresponding Fourier transforms (**b**)

evaluated. A coupling efficiency of 80–85 % was observed up to 30 W of incident power before damage to the fibre occurred. It is interesting that this damage was not instantaneous and occurred after approximately 1 min of operation. This may be assigned to the influence of thermal effects or cumulative effects in the fibre. At a lower incident power of 25 W the fibre could be used for many minutes with no sign of damage. The spectral broadening corresponding to this power level is shown in Fig. 6.2a. The spectrum spans the range 925–1,140 nm and shows a periodic modulation characteristic for SPM. The Fourier transform limit of this spectrum on the assumption of a flat phase in Fig. 6.2b is 17 fs. In order to increase the coupling energy a 4.5-cm-long LMA 35 fibre with larger core diameter was chosen. It was possible to couple in the whole power of 45 W available from the oscillator with a coupling efficiency of 90–92 %, limited by the light reflection at the facets of the fibre. Such high efficiency was possible due to the larger core diameter and excellent mode quality ($M^2 = 1.1$) of the oscillator. No damage or degradation was noticed after several minutes of operation. It is good to note that the oscillator with pulse energy Ep = 1.1 μJ and pulse duration $\tau = 250$ fs has a peak power of P = 3.9 MW, which is as high as the critical peak power for self-focusing in fused silica $P_{cr} =$ 4.3 MW (see Eq. 5.4). Thus it is still possible to launch the whole pulse energy into the fibre and work near the point of critical self-focusing. This spectral broadening corresponding to the maximal power level is shown in Fig. 6.2. The spectrum spans a slightly narrower range than that achieved with the LMA 25, which can be attributed to the shorter fibre length and accuracy of the measurements (measured with a single-mode fibre and Ando AQ6315). The Fourier transform-limit of this spectrum corresponds to 19 fs. Further investigations on degradation or fibre damage at maximal input power or optimization of the length with regard to the maximal achievable spectral broadening were not performed, because the dispersive mirrors available at the time of the experiment supported only bandwidths from 980 to 1,080 nm, as shown in Fig. 6.3. Thus the optimal spectral broadening which fits to the bandwidth of these mirrors was obtained at ~17–19 W input power. Compression of these pulses

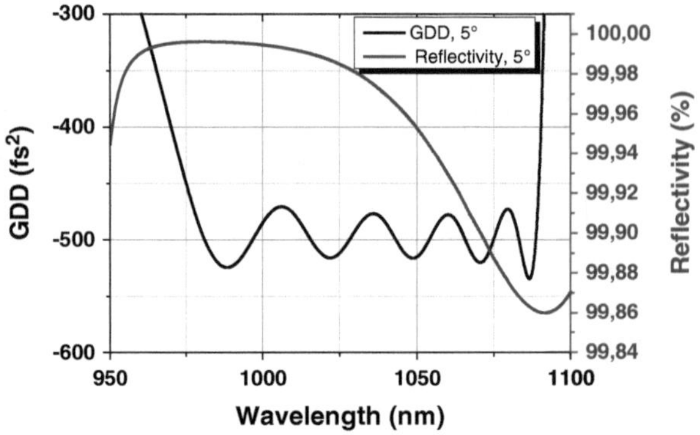

Fig. 6.3 Theoretical GDD and reflectivity versus wavelength of CM 39 dispersive mirrors used for pulse compression experiments

was accomplished by 6 bounces on two mirrors CM 39[2] with GDD $= -500\, fs^2$ and without compensation of higher-order dispersion, as shown in Fig. 6.4. Autocorrelation measurement yields 60 fs, which on the assumption of sech^2 pulses and the corresponding deconvolution factor of 0.65 results in 40-fs pulse duration. This value is slightly larger than suggested by the spectral Fourier transform of 32 fs, which can be explained by a non-optimized compressor. In particular, neither fine tuning of the GDD nor careful alignment of the incident angles on the dispersive mirrors was done. The nonlinear part of the chirp could also contribute. Coupling of higher powers into the fibre resulted in a broader spectrum and the appearance of a pronounced pedestal in the autocorrelation measurement. The pulse duration of 40 fs was short enough to perform further experiments on CE phase detection. It is interesting to estimate the characteristic nonlinear and dispersion lengths L_{NL} and L_D for this case. From [11] the nonlinear coefficient $\gamma = 0.7\,\text{W}^{-1}\text{km}^{-1}$ and the dispersion coefficient $\beta_2 = 0.29$ ps^2/km[3] can be taken. Substituting these values and oscillator parameters in Eq. (6.1) yields $L_{NL} = 0.4\,\text{mm}$ and $L_D = 2.2\,\text{m}$. By comparing these length scales with the length of our fibre (60 mm), it can be concluded that nonlinear effects predominate and dispersive effects can reasonably be neglected.

The experiments on spectral broadening with the thin-disk laser in [6, 12] showed a certain degradation of polarization which can be attributed to nonlinear polarization rotation in the fibre or its slight asymmetry. In contrast, spectral broadening with a Ti:S oscillator showed no degradation of the polarization extinction ratio [12]. Measurement of the polarization after spectral broadening was not carried out in this experiment and is a subject for further work.

[2] designed and manufactured by V. Pervak

[3] the coefficients defined for the LMA 25 fibre

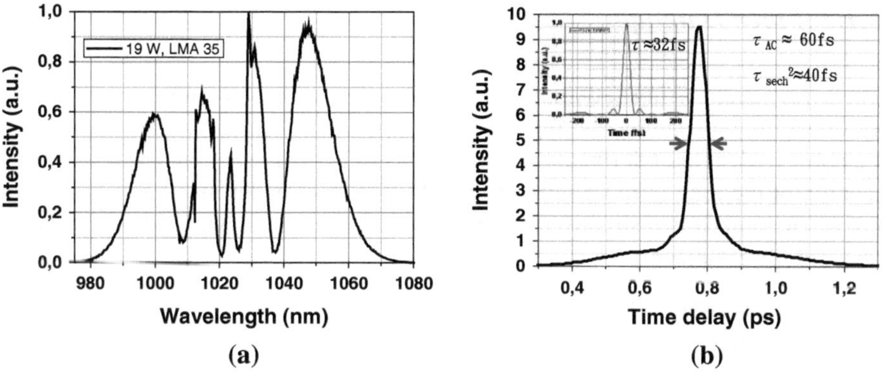

Fig. 6.4 Spectral broadening obtained in LMA 35 fibre with 19 W (**a**) and corresponding autocorrelation measurement (with APE pulse check autocorrelator) (**b**) *Inset* shows fourier transform of the left spectrum

6.1.2 Two-Stage Fibre Compression: Vision

Let us assume that the spectrum in Fig. 6.2a is compressible down to 20 fs, which is slightly above its Fourier limit in Fig. 6.2b. The overall losses of the compressor designed to support this bandwidth can be as low as 5–10 %. Together with fibre transmission losses of 10 % the whole setup will be able to deliver 0.9 μJ and 20 fs, which correspond to ~40-MW peak power. With further optimization even the sub-15 fs range can be accessed. Additionally, one could consider the realizing a second spectral broadening stage by implementing the concept of chirped-pulse compression introduced in [10]. Pulses of 55 fs and 0.36 μJ were compressed down to 14.5 fs by positively chirping them before the fibre [10] and thus reducing the peak power below critical. The most attractive outcome was found in chirp transfer. In particular, only the nineth part of the positive GDD needed for chirping had to be compensated by the negative GDD in the compressor [12]. This value was almost constant in the range from 55 fs up to positively chirped 250-fs pulses. It can be partially explained by assuming that the chirp spreads over a large spectral bandwidth and thus has smaller GDD values to compensate. In practice, these findings greatly reduce demands on the amount of GDD in a compressor (second compressor in this case). It is worth thinking about the optimal value of pulse duration after the first-stage compressor in order to achieve few-cycle pulses in a second stage and still have low overall losses (<20 %) in both compressors. The more complex and time-consuming alignment can be assigned to drawbacks of such a two-stage spectral broadening approach. Preliminary experiments made by Fabian Lücking showed promising results on accessing close to octave-spanning spectral broadening after the second stage.

6.1.3 Fibre Damage

The experiments on fibre compression down to 40 fs with 19-W input power showed excellent output stability and reproducibility on a daily basis. No fibre degradation was observed, at least during the several weeks of experimentation. However, investigation of the long-term stability of the LMA 35 fibre under maximum power was not carried out. It is instructive to summarize the measures, which should be taken to avoid fibre damage.

- remove the plastic coating
- avoid strong focusing on the facet of the fibre
- avoid fibre misalignment due to thermal effects
- keep the peak power below the self-focusing threshold
- avoid fibre contamination by dust or liquids
- use end caps and tapered fibres (not used in this work)

6.1.4 Further Steps

Pulse energies of thin-disk oscillators [13, 14], fibre and innoslab amplifier laser systems [15, 16] approach 100-μJ level at MHz repetition rate. Due to self-focusing solid-core PCFs are limited to approximately 4 MW in peak power, thus only several μJ with sub-ps pulse duration can be compressed with these fibres. Chirped pulse compression can be considered as a way to increase this limit, but strong chirping of the input pulses to several picoseconds is rather undesirable because of the constraints on the compressor optics. The best solution lies in implementing spectral broadening in hollow-core PCF filled with a noble gas at high pressure [17]. Of particular interest are, kagomé PCF with pressure-tunable dispersion, tolerating high guided intensities over 10^{14} W/cm^2. The proof-of-principle experiments on spectral broadening with a kagomé fibre and the TD oscillator were performed in [18]. A detailed analysis with simulations for input energies $>10\,\mu$J can be found in review paper [19]. The handling of high average powers in fibres may also be a limiting factor analogous to the thermal instabilities observed in CPA fibre lasers [17].

Optimization of the maximal spectral broadening, polarization properties and implementation of the two-stage compression is subject to further work. The kagomé-based fibre compression is considered as a promising candidate in the near future for experiments with an energy-upscaled version of the described KLM Yb:YAG oscillator.

6.2 CE Phase Measurement and Stabilization

It is interesting to remember the experiment of Gohle et al. [20] in 2005 when femtosecond pulses from a Ti:S oscillator were externally enhanced in a passive cavity up to 28 fs and 38 W. As shown earlier, our system has very similar characteristics, but with output directly from the oscillator and with fibre broadening in only one stage. To make it more usable for attosecond and frequency metrology CE phase stabilization is needed.

6.2.1 CE Phase and CE Frequency

As shown in Fig. 6.5, the pulse train generated by a mode-locked oscillator corresponds to equidistantly-spaced frequency components which are in phase (locked) and form a frequency comb. In the time domain the difference between the maximum of the envelope and the closest carrier maximum corresponds to the carrier envelope phase $\Delta\varphi$ as shown in Fig. 6.5. Moreover the free-running oscillator exhibits a change of the CE phase $\Delta\varphi$ from pulse to pulse. This change becomes very

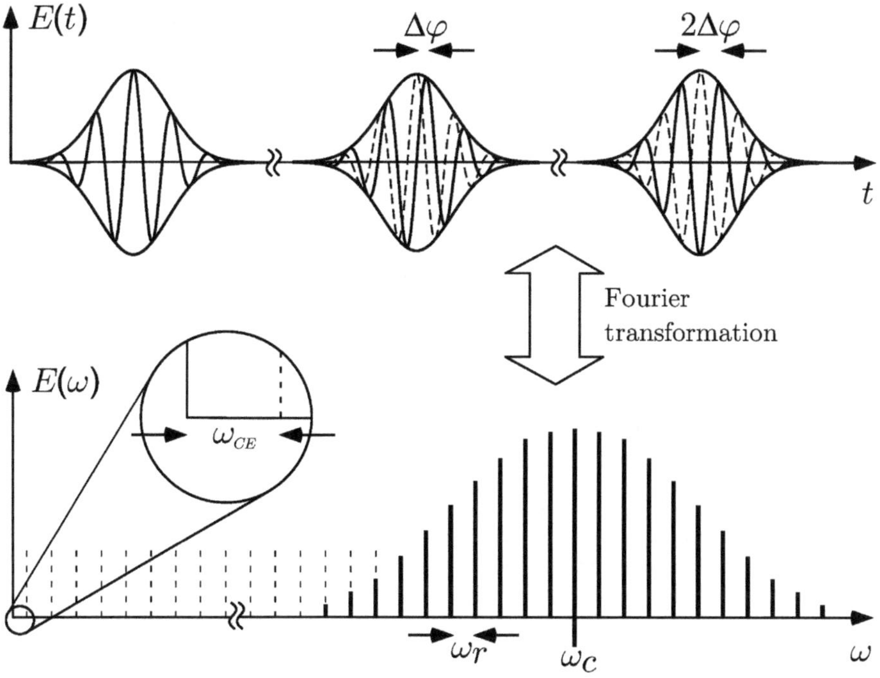

Fig. 6.5 Pulse train emitted by a mode-locked oscillator (*top*) and frequency comb with a frequency offset ω_{CE}. Picture taken from [21]

important when pulses approach the few-cycle regime and their electric field starts to depend distinctly on the CE phase. Strong field experiments such as HHG [1] need to have a controllable and stable CE phase with only minimal fluctuations in order to maximize the electric field on the target. In the frequency domain a shift of the whole comb corresponds to the carrier envelope offset frequency f_{CEO}, which can be expressed via $\Delta\varphi$ as

$$f_{CEO} = f_r \frac{\Delta\varphi}{2\pi}.$$

Moreover, stabilization of f_{CEO} fixes the position of the whole comb in frequency space:

$$f_n = nf_r + f_{CEO},$$

where f_n and f_r are one of the comb lines and the repetition rate of the oscillator, respectively. This relation links optical frequency f_n to radio frequencies f_r and f_{CEO}. The latter has a drastic impact on frequency metrology [22]. Determination of f_{CEO}, however, is not as trivial as the measurement of f_r, which can simply be done with a photodiode and frequency counter. f_{CEO} frequency measurement can be accomplished by having an octave-spanning spectrum from the oscillator and heterodyning modes obtained via nonlinear frequency conversion. In particular, taking a high-frequency component at $f_{2n} = 2nf_r + f_{CEO}$ and heterodyning it with a second-harmonic low-frequency f_n component, i.e. $2f_n = 2nf_r + 2f_{CEO}$ yields

$$2f_n - f_{2n} = 2nf_r + 2f_{CEO} - 2nf_r - f_{CEO} = f_{CEO}. \tag{6.2}$$

This scheme was implemented for the first time in [23, 24] by generating octave-spanning spectra with 100-fs pulses in a small-core PCF [25] or in a short piece of a telecom single-mode fibre seeded by a sub-10-fs Ti:S oscillator [26]. The details on this scheme will be presented further on by considering the specific case of our TD oscillator.

6.2.2 Generation of One-Octave Spectrum

In contrast to spectral broadening experiments shown at the beginning of this chapter when PCF with large core diameters and positive dispersion are used, the use of small core diameter <3 μm PCF with negative (or zero) dispersion at the central wavelength of the launched soliton allows one to generate a supercontinuum (see review [27]) with only small input energy (thanks to the high intensities over a long propagation distance). This generation process is highly nonlinear and causes soliton fission [27]. The supercontinuum spectrum can maintain coherence for short seeding pulse durations (typically <100 fs) and thus enable detection of f_{CEO} as in Eq. (6.2).

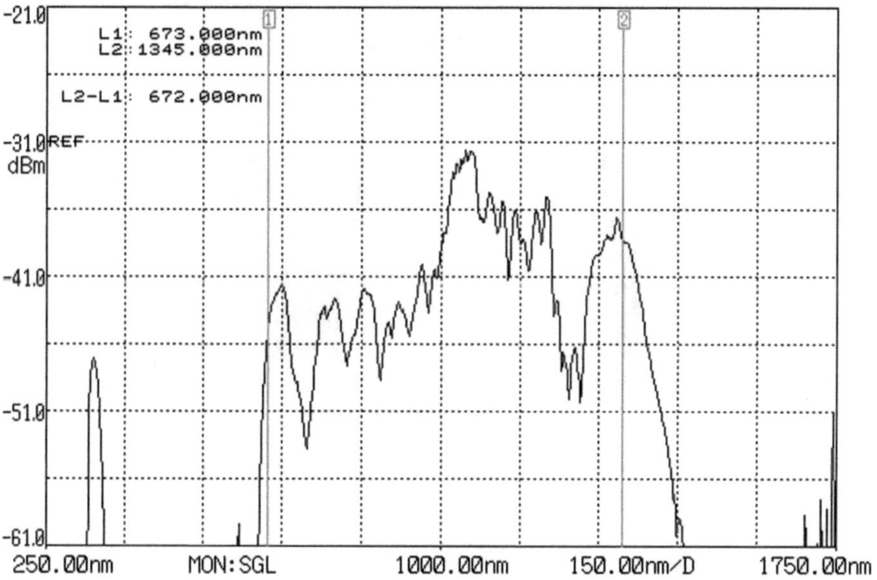

Fig. 6.6 An octave spanning spectrum broadened in PCF (SC-3.7-975, NKT photonics) by seeding it with <500 mW, 40 fs pulses

The spectrum generated in our lab in PCF (SC-3.7-975, NKT photonics) with 3.7-μm core diameter by launching <13 nJ and 40 fs pulses is shown in Fig. 6.6. The spectrum covers more than one octave and even broader spectra could be generated by sending slightly higher pulse energies. One of the serious drawbacks of such an approach to generate a continuum is the very alignment-sensitive focusing into the small-core PCF, thus limiting long-term CE phase stability. It is worth noting that the longer the initial pulses are, the longer is the fibre necessary to generate the continuum, which leads to excessive noise.

6.3 Octave-Spanning Spectrum in Bulk Crystals

White-light generation in bulk materials can be considered as an interesting alternative to supercontinuum generation in PCF with reduced constraints on laser beam pointing. Moreover, the spectrum generated in bulk crystals can be compressed temporally (in contrast to the spectrum in Fig. 6.6 after the PCF), which in turn can allow realization of a compact NOPA fully based [7, 28] on a TD oscillator. A comprehensive overview of white-light generation in different laser host crystals is presented in [29]. Here, the preliminary experiment on white light generated in a YAG crystal is described. The compressed oscillator output of <6 W and 40 fs was focused into a 4-mm-thick YAG crystal by means of a 100-mm lens. These parameters corresponded

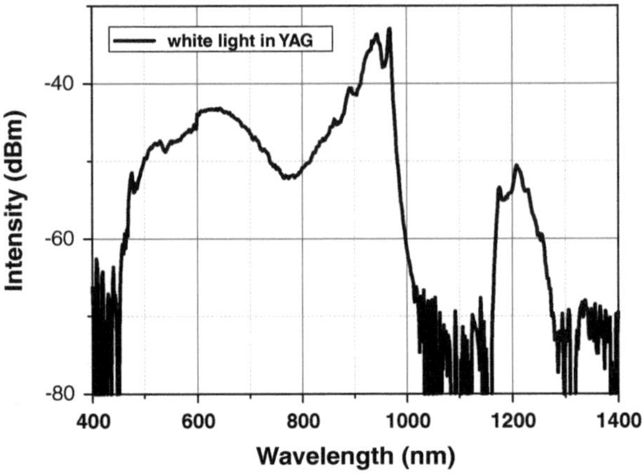

Fig. 6.7 White-light generation in YAG crystal. Spectral dips at 1,030–1,300 nm are due to filtering by the HR mirror

to the approximate threshold value for white-light generation. An HR mirror was placed right after the crystal to filter out most of the radiation centred at 1,030 nm. The spectrum measured after this HR mirror is shown in Fig. 6.7 and indicates spectral broadening over two octaves. The dips near 1,030–1,300 nm are caused by the filtering action of the HR mirror. This preliminary experiment shows great potential for using a continuum generated in bulk YAG for both derivation of a beat note f_{CEO} and generation of a seed pulse for NOPA. On the other hand, the low conversion efficiency and necessity of tight focusing can be considered as drawbacks. For these reasons and a better understanding of PCF-based continuum generation, the latter was chosen for further experiments.

6.3.1 CE Phase Measurement

Heterodyning of two spectral components f_{2n} and $2f_n$ of an octave-spanning frequency comb implies their spatial and temporal overlapping. This can be accomplished by means of a so-called f-2f interferometer [26, 23] as depicted in Fig. 6.8. Fibre broadening in the first stage provides pulses as short as 40 fs at up to an output power of 40 W. Afterwards this power can be attenuated continuously in the range 300–700 mW by an arrangement consisting of a wedge and a half-wave plate. The beam is then focused into a 3.7-μm-diameter-core PCF by an aspheric lens with focal distance f = 8 mm (C240TM, Thorlabs) and collimated by a lens with f = 4.5 mm (C230TM, Thorlabs). An input power of ~500 mW, corresponding to ~13 nJ, is typically enough to generate over one-octave-broad spectra as shown in Fig. 6.6. The collimated beam is sent through a prism sequence, which allow separation of infrared and visible (red) portions of the continuum [30]. These portions are separately

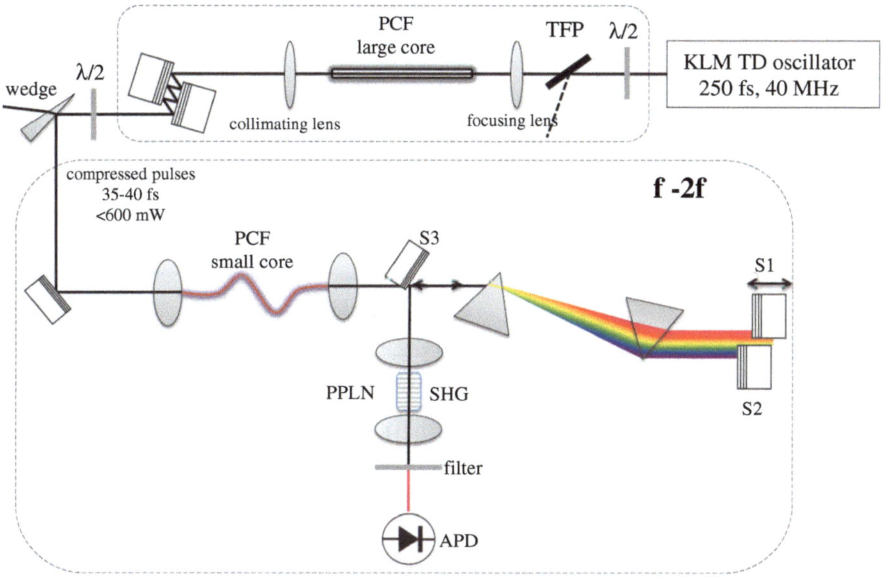

Fig. 6.8 The setup for CE offset frequency measurement

reflected back by two mirrors *S1* and *S2*. The delay introduced by the mirror *S1* is <2 mm and is necessary to compensate for the delay introduced by the PCF during continuum generation. Retroreflection is done with a slight misalignment in the vertical direction in order to hit mirror *S3* and direct the beam to the frequency-doubling stage. The latter consists of a periodically-poled lithium niobate crystal (PPLN) placed between the focusing and collimating lenses. PPLN (Covesion 1375-1) is optimized to frequency-double the infrared part of the spectrum near ∼1,375 nm. The generated ∼687-nm light is band-passed (filter FB680-10-1) and detected by an avalanche photodiode (APD 210, Menlosystems). A measured beat signal is shown in Fig. 6.9. Two things are noticeable: the good signal-to-noise ratio (SNR)>35 dBm and the pronounced fluctuations of f_{CEO} within the ∼2 MHz range. However, the signal stays stable in amplitude for several hours, with no need to align either the first or the second stage with PCF.

6.4 CE Phase Noise and Stabilization

The physical origin of CE phase variations from pulse to pulse can be understood by considering the difference between the group- υ_g and phase- υ_p velocities of pulses propagating through a medium of length L [31]:

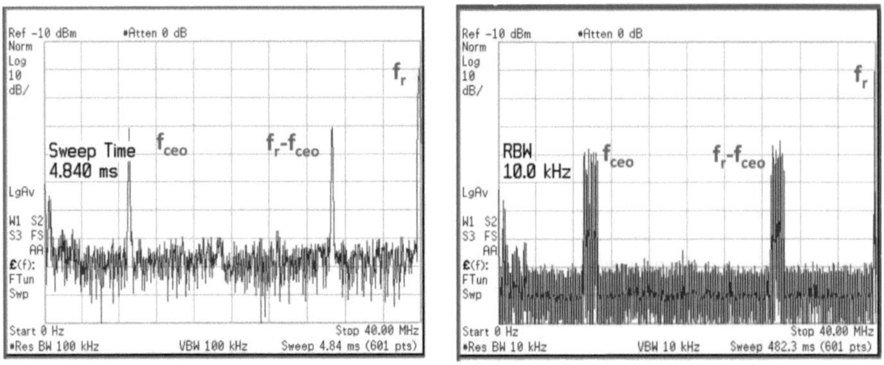

(a) **(b)**

Fig. 6.9 Beat signal measured at 100 kHz (**a**) and 10 kHz (**b**) resolution bandwidth. Picture **a** shows CE frequency fluctuations within the ~2 MHz bandwidth with SNR>35 dBm

$$\triangle\varphi = 2\pi f_c L(\frac{1}{v_p} - \frac{1}{v_g}) = 2\pi L(\frac{dn}{d\lambda})|_{\lambda_0}. \tag{6.3}$$

The difference in the CE phase shift in Eq. (6.3) is only due to the material dispersion n(λ) and can be classified as a linear contribution. Nonlinear contributions due to the Kerr and Raman effects are not only possible but are frequently used in stabilization of the CE phase [23, 32]. A change of the pump current varies the intracavity peak power, which in turn influences the CE phase shift. In order to estimate the demands on the CE phase stabilization scheme, the CE phase sensitivity to the variation of the pump current was investigated for our oscillator. The oscillator shows a high pump sensitivity of ~4 W/40 MHz at 200-W pump power. Thus f_{CEO} can span the whole range from 0 to f_r (40 MHz) when the pump power is changed by <2 %. In contrast to the experiments with Ti:S oscillators, there is no need for an additional f_{CEO} coarse-tuning mechanism as has been done with, for example, a couple of wedges [26]. On the other hand, such high sensitivity means a strong dependence of CE phase on pump power fluctuations. According to the specification of the supplier the power fluctuations of the pump diodes are ~1 % peak-to-peak value. Thus, one origin of the CE phase fluctuations can clearly be identified as the noise of pump diodes. The second source of noise is likely to be the TD cooling geometry (see Chap. 3), which implies high water flow hitting the 1-mm-thick CuW heat sink to which the TD is mounted. As was noted in Chap. 4, this has an impact on the oscillator intensity stability and thus the cooling flow was minimized to 0.5 l/min, which is 4 times less than recommended by manufacture. Both noise of the pump diodes and acoustic noise induced by the active cooling are the most likely contributions to the f_{CEO} jitter. This technical noise lies typically at <100 kHz. The radio frequency spectrum in Fig. 5.13a is in agreement with this estimation. The pump diodes available for the experiment, however, had an amplitude response bandwidth <1 kHz as preliminary measurements show. There are other mechanisms leading to the extra noise: temperature and air flow fluctuations and proximity to the QML threshold.

6.4.1 CE Phase-Locking

There are different schemes of CE phase stabilization involving the intra- or extra-cavity use of an AOM or EOM [23] for power modulation of an oscillator, the swivel-piezo-method for an oscillator containing a prism pair [24] and the use of an extra-cavity AO frequency shifter [33, 34]. Another way of stabilizing the CE phase is to control the current of the laser diodes via a phase-locked loop (PLL) [35, 36]. Even though the modulation bandwidth of our pump diodes is rather narrow, a proof-of-principle experiment on the stabilization of f_{CEO} was performed. Details of the locking scheme are not described here as this scheme is rather standard and utilizes a large-range digital phase detector and PI controller (LB1005, Precision Photonics). The error signal from the PI controller was fed to the modulation input of the pump diodes. The f_{CEO} was tuned to stay close to 10 MHz frequency, then this signal was band-passed, amplified and sent to one of the inputs of the phase detector. The reference signal was obtained from a stable RF generator and fed to the second input of the phase detector. Only negligible locking of the CE frequency could be achieved. This locking mainly indicated the fact that f_{CEO} can be pulled towards the reference frequency by the PLL and stay near that frequency, but however showing similar fluctuations to those for the free-running case (see Fig. 6.9).

In order to understand the reason for such a poor lock, one has to make a systematic and precise measurement of the frequency-dependent amplitude and phase response of the diode laser power, the oscillator output power and the f_{CEO} to modulation of the diode current. In [35] these characteristics were measured for an Yb:KYW SESAM-ML bulk oscillator at 160 MHz. It is interesting that the amplitude response showed a roll-off at ~20–30 kHz. The fluorescence lifetime of Yb:KYW is ~0.3 ms, which is 3 times less than that for Yb:YAG (see Table 3.2). As follows from this rough comparison, the modulation bandwidth, limited by the gain dynamics should be in the range 7–10 kHz for our oscillator. Very likely the phase response of the pump diodes used limits the bandwidth well below 1 kHz. It is worth noting that CE frequency was recently measured from a low-power SESAM-ML $Yb : Lu_2O_3$ TD oscillator in [37] . However, extracting any details about the CE frequency noise and its origins is not possible from [37].

6.5 Further Steps

Spectral broadening and compression of 250 fs, 1.1 μJ pulses from the TD oscillator resulted in 40-fs pulses, limited by the bandwidth of the available chirped mirrors. Spectral broadening with a corresponding Fourier limit of 19 fs was achieved at full power throughput of 40 W. Further compression experiments and optimization should result in ~15-fs pulses at around 30 W average power. Two-stage spectral broadening in PCF together with chirped pulse compression [10] seems a promising way of reaching sub-10-fs (3-cycle) pulses.

The f_{CEO} beat signal from a KLM TD oscillator was derived for the first time. Preliminary experiments on stabilization and characterization of f_{CEO} show that eliminating noise from the pump diodes as well as acoustic noise from the active TD cooling should allow CE phase-locking with at least sub-radian stability.

References

1. T. Brabec, F. Krausz, Intense few-cycle laser fields: frontiers of nonlinear optics. Rev. Mod. Phys. **72**, 545–591 (2000)
2. G.P. Agrawal, *Nonlinear Fiber Optics*, 3rd edn. (Academic Press, New york, 2001)
3. W.J. Tomlinson, R.H. Stolen, C.V. Shank, Compression of optical pulses chirped by self-phase modulation in fibers. J. Opt. Soc. Am. B **1**(2), 139–149 (1984)
4. T.A. Birks, J.C. Knight, P.S. Russell, Endlessly single-mode photonic crystal fiber. Opt. Lett. **22**(13), 961–963 (1997)
5. P. Russell, Photonic crystal fibers. Science **299**(5605), 358–362 (2003)
6. T. Südmeyer, F. Brunner, E. Innerhofer, R. Paschotta, K. Furusawa, J.C. Baggett, T.M. Monro, D.J. Richardson, U. Keller, Nonlinear femtosecond pulse compression at high average power levels by use of a large-mode-area holey fiber. Opt. Lett. **28**(20), 1951–1953 (2003)
7. A. Steinmann, A. Killi, G. Palmer, T. Binhammer, U. Morgner, Generation of few-cycle pulses directly from a MHz-NOPA. Opt. Express **14**(22), 10627–10630 (2006)
8. T. Eidam, F. Röser, O. Schmidt, J. Limpert, A. Tünnermann, 57 W, 27 fs pulses from a fiber laser system using nonlinear compression. Appl. Phys. B **92**, 9–12 (2008)
9. A. Vernaleken, J. Weitenberg, T. Sartorius, P. Russbueldt, W. Schneider, S.L. Stebbings, M.F. Kling, P. Hommelhoff, H.-D. Hoffmann, R. Poprawe, F. Krausz, T.W. Hänsch, T. Udem, Single-pass high-harmonic generation at 20.8 Mhz repetition rate. Opt. Lett. **36**(17), 3428–3430 (2011)
10. T. Ganz, V. Pervak, A. Apolonski, P. Baum, 16 fs, 350 nJ pulses at 5 MHz repetition rate delivered by chirped pulse compression in fibers. Opt. Lett. **36**(7), 1107–1109 (2011)
11. R. Kuis, A. Johnson, S. Trivedi, Measurement of the effective nonlinear and dispersion coefficients in optical fibers by the induced grating autocorrelation technique. Opt. Express **19**(3), 1755–1766 (2011)
12. T. Ganz, *Chirped Pulse Compression for High Energy Laser Systems at MHz Repetition Rates*, Ph.D. thesis, LMU München, 2011
13. D. Bauer, I. Zawischa, D.H. Sutter, A. Killi, T. Dekorsy, Mode-locked Yb:YAG thin-disk oscillator with 41 μJ pulse energy at 145 W average infrared power and high power frequency conversion. Opt. Express **20**, 9698 (2012)
14. S.V. Marchese, C.R. Baer, A.G. Engqvist, S. Hashimoto, D.J. Maas, M. Golling, T. Südmeyer, U. Keller, Femtosecond thin disk laser oscillator with pulse energy beyond the 10-microjoule level. Opt. Express **16**(9), 6397–6407 (2008)
15. T. Eidam, S. Hanf, E. Seise, T.V. Andersen, T. Gabler, C. Wirth, T. Schreiber, J. Limpert, A. Tünnermann, Femtosecond fiber CPA system emitting 830 W average output power. Opt. Lett. **35**(2), 94–96 (2010)
16. P. Russbueldt, T. Mans, J. Weitenberg, H.D. Hoffmann, R. Poprawe, Compact diode-pumped 1.1 kW Yb:YAG innoslab femtosecond amplifier. Opt. Lett. **35**(24), 4169–4171 (2010)
17. C. Jauregui, T. Eidam, H.-J. Otto, F. Stutzki, F. Jansen, J. Limpert, A. Tünnermann, Physical origin of mode instabilities in high-power fiber laser systems. Opt. Express **20**(12), 12912–12925 (2012)

18. O.H. Heckl, C.J. Saraceno, C.R.E. Baer, T. Südmeyer, Y.Y. Wang, Y. Cheng, F. Benabid, U. Keller, Temporal pulse compression in a xenon-filled Kagome-type hollow-core photonic crystal fiber at high average power. Opt. Express **19**, 19142–19149 (2011)
19. J.C. Travers, W. Chang, J. Nold, N.Y. Joly, P.S.J. Russell, Ultrafast nonlinear optics in gas-filled hollow-core photonic crystal fibers. J. Opt. Soc. Am. B **28**(12), A11–A26 (2011)
20. C. Gohle, T.U.M. Herrmann, J. Rauschenberger, R. Holzwarth, H.A. Schuessler, F. Krausz, T.W. Hänsch, A frequency comb in the extreme ultraviolet. Nature **436**, 234–237 (2005)
21. T. Udem, R. Holzwarth, M. Zimmermann, C. Gohle, T. Hänsch, in *Few-Cycle Laser Pulse Generation and Its Applications*, Optical Frequency-Comb Generation and High-Resolution Laser Spectroscopy, (Springer, Berlin, 2004), pp. 295–313
22. T. Udem, R. Holzwarth, T.W. Hänsch, Optical frequency metrology. Nature **416**, 233–237 (2002)
23. R. Holzwarth, T. Udem, T.W. Hänsch, J.C. Knight, W.J. Wadsworth, P.S.J. Russell, Optical frequency synthesizer for precision spectroscopy. Phys. Rev. Lett. **85**, 2264–2267 (2000)
24. D.J. Jones, S.A. Diddams, J.K. Ranka, A. Stentz, R.S. Windeler, J.L. Hall, S.T. Cundiff, Carrier-envelope phase control of femtosecond mode-locked lasers and direct optical frequency synthesis. Science **288**(5466), 635–639 (2000)
25. J.C. Knight, T.A. Birks, P.S.J. Russell, D.M. Atkin, All-silica single-mode optical fiber with photonic crystal cladding. Opt. Lett. **21**(19), 1547–1549 (1996)
26. A. Apolonski, A. Poppe, G. Tempea, C. Spielmann, T. Udem, R. Holzwarth, T.W. Hänsch, F. Krausz, Controlling the phase evolution of few-cycle light pulses. Phys. Rev. Lett. **85**, 740–743 (2000)
27. J.M. Dudley, G. Genty, S. Coen, Supercontinuum generation in photonic crystal fiber. Rev. Mod. Phys. **78**, 1135–1184 (2006)
28. H. Fattahi, C.Y. Teisset, O. Pronin, A. Sugita, R. Graf, V. Pervak, X. Gu, T. Metzger, Z. Major, F. Krausz, A. Apolonski, Pump-seed synchronization for MHz repetition rate, high-power optical parametric chirped pulse amplification. Opt. Express **20**(9), 9833–9840 (2012)
29. M. Bradler, P. Baum, E. Riedle, Femtosecond continuum generation in bulk laser host materials with sub-μJ pump pulses. Appl. Phys. B **97**, 561–574 (2009)
30. C. Grebing, S. Koke, B. Manschwetus, G. Steinmeyer, Common-path interferometer for incorruptible detection of the carrier-envelope phase drift, in *Conference on Quantum Electronics and Laser Science. CLEO/QELS 2008*, May 2008, pp. 1–2
31. G. Tempea, R. Holzwarth, A. Apolonski, T. Hänsch, F. Krausz, in *Ultrafast Lasers: Technology and Applications*, Phase-Controlled Few-Cycle Light, (Marcel Dekker Inc, New York) pp. 573–610 (2002)
32. A. Poppe, R. Holzwarth, A. Apolonski, G. Tempea, C. Spielmann, T. Hänsch, F. Krausz, Few-cycle optical waveform synthesis. Appl. Phys. B **72**, 373–376 (2001)
33. S. Koke, C. Grebing, H. Frei, A. Anderson, A. Assion, G. Steinmeyer, Direct frequency comb synthesis with arbitrary offset and shot-noise-limited phase noise. Nature Photon. **4**, 462–465 (2010)
34. F. Lücking, A. Assion, A. Apolonski, F. Krausz, G. Steinmeyer, Long-term carrier-envelope-phase-stable few-cycle pulses by use of the feed-forward method. Opt. Lett. **37**(11), 2076–2078 (2012)
35. S.A. Meyer, J.A. Squier, S.A. Diddams, Diode-pumped Yb:KYW femtosecond laser frequency comb with stabilized carrier-envelope offset frequency. Eur. Phys. J. D. **48**(1), 19–26 (2008)
36. B.R. Washburn, W.C. Swann, N.R. Newbury, Response dynamics of the frequency comb output from a femtosecond fiber laser. Opt. Express **13**(26), 10622–10633 (2005)
37. C.J. Saraceno, S. Pekarek, O.H. Heckl, C.R.E. Baer, C. Schriber, M. Golling, K. Beil, C. Kränkel, G. Huber, U. Keller, T. Südmeyer, Self-referenceable frequency comb from an ultrafast thin disk laser. Opt. Express **20**(9), 9650–9656 (2012)

Chapter 7
XUV Output Coupler and XUV/IR Grazing-Incidence Beam Splitter

As discussed in the introduction chapter, several scenarios for driving nonlinear processes are possible: directly by an oscillator, inside an oscillator or inside a passive resonator and directly by amplified oscillator output. In most cases the XUV and the driving field have to be coupled out from a resonator and/or separated for further experiments. Separation/outcoupling of the XUV and driving field inside the enhancement cavity is the most demanding and richest in constraints among these different approaches. Finding a solution for this problem at extreme conditions allows one to transfer this solution to other XUV-generation approaches, but with strongly relaxed constraints. Such a solution is found in a so-called grazing-incidence plate (GIP).

State-of-the-art femtosecond enhancement cavities and demands on an XUV output coupler are presented in Sects. 7.1 and 7.2, respectively. An overview of different XUV output couplers is introduced in Sect. 7.3. Section 7.4 describes in detail the concept, technical realization and characterization in the XUV and NIR ranges of GIP. It also includes a discussion of the advantages and disadvantages of the GIP. The applicability of GIP for intra- and extra-cavity XUV experiments is shown in Sect. 7.5. A brief conclusion summarizes this chapter. This chapter comprises material previously published in [1].

7.1 Femtosecond Enhancement Cavity

The use of a high-finesse optical cavity for coherent storage of radiation is a commonly used technique for efficient frequency conversion of CW lasers [2]. This technique was recently extended to mode-locked lasers [3] and has now become widespread in research groups throughout the world [4, 5]. By using the enhancement cavity approach, one can greatly increase the power from a seed mode-locked laser oscillator inside the cavity (the enhancement factor is typically 10–10^4). The conditions necessary to realize a large enhancement factor and high average power include the use of high-damage-threshold, thermally-stable intracavity optics with

O. Pronin, *Towards a Compact Thin-Disk-Based Femtosecond XUV Source*, Springer Theses, DOI: 10.1007/978-3-319-01511-8_7,

extremely low losses, as well as very fine dispersion control of the cavity. Clearly the enhancement cavity has to be kept in resonance with a seeding oscillator and mode-matched to it. This is accomplished via several feedback loops which control both the resonator length and CE phase [6].

The first generation of enhancement cavities relied on Ti:sapphire lasers as seeding source, resulting in circulating intracavity pulses of 28 and 60-fs duration and 38 and 480-W average power, respectively [4, 5]. Recent development and power scaling of enhancement cavities with Yb-based fibre amplifiers [7] as seeding source approached the 7-kW level for a cavity with an XUV output coupler in it [8, 9] and more than 18 kW for the empty cavity [10]. More advanced cavity designs were proposed to avoid current limitations in further power scaling [11]. At the moment, intracavity-based high-harmonic generation is the most promising way of approaching power-scalable compact and coherent XUV MHz-repetition-rate sources. However, not only can a passive enhancement cavity be considered for increasing the laser power. The possibility of utilizing high average powers inside the laser oscillator cavity has become a reality with recent progress in high-power femtosecond thin-disk lasers. Typically, the output coupler transmission T of an oscillator amounts to only a few per cent. This means that power stored inside the oscillator cavity is a factor 1/T higher than the output power. Very recently, a first proof-of-principle experiment on XUV generation inside a Ti:Sa oscillator was realized by Seres et al. [12].

7.2 Demands on the XUV Output Coupler

When XUV radiation is generated inside the cavity, its output coupling immediately presents a challenge. XUV light is generated collinearly with the driving fundamental laser beam (see Fig. 7.1) and can easily be absorbed even by 1-μm-thick condensed matter, e.g. by the multilayer structure of a mirror, which typically exhibits very poor reflectance of the order of $\sim 10^{-4}$ at normal incidence [13].

In the following, the necessary conditions for an intracavity XUV output coupler are summarized:

(a) High XUV reflectivity. All XUV light power generated inside the cavity should ideally be coupled out.
(b) Broad range of XUV reflectance. All generated harmonics of the fundamental of DF should be coupled out collinearly.
(c) Losses introduced by the OC for the DF should be low.
(d) The dispersion introduced by the OC should be small. Nonlinear effects introducing intensity-dependent GDD should be small.
(e) Low thermal lensing. High thermal conductivity and low thermal expansion, low thermo-optical coefficient are necessary to avoid power-dependent resonator stability behaviour.
(f) High damage threshold.

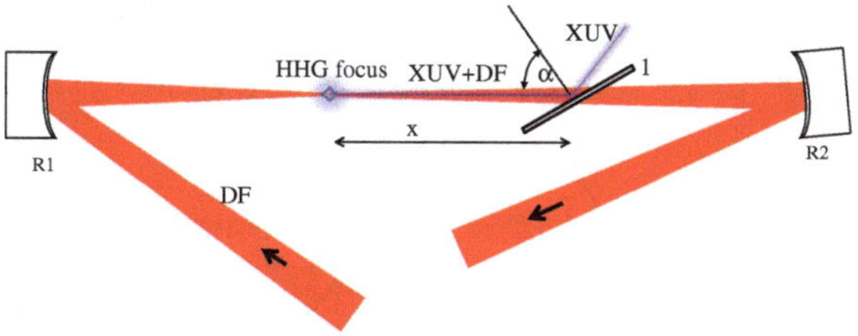

Fig. 7.1 Brewster plate (1) inside the ring enhancement cavity (shown, only 2 concave mirrors *R1, R2* of the cavity). In this case α is equal to the Brewster angle affording lowest losses for the circulating DF of p-polarized light. In one of the foci of the cavity (HHG focus), a gas medium (usually a jet) is placed for generating harmonics of DF. Then, XUV and DF co-propagate towards the Brewster plate, where XUV becomes partially reflected out of the cavity, whereas DF propagates through the plate without losses and beam distortion [1]

Moreover, damage and degradation of different XUV output couplers under UV/XUV irradiation was observed in [8, 14].

7.3 Overview of Existing XUV Output Couplers and Methods

7.3.1 Brewster Plate

The simplest XUV OC demonstrated so far consists of a plate of material with good optical quality that is transparent to the DF and is placed at the Brewster angle of incidence inside the enhancement cavity between the focus and concave mirror, see Fig. 7.1. In this case losses of the p-polarized DF are nearly zero. Due to the difference of the refractive indices of XUV and DF, a small reflection of XUV occurs at the surface of the plate. The reflection of sapphire and fused silica plates, used in the previous experiments [4, 5, 15], for p-polarized XUV light at the Brewster angle for the DF is shown in Fig. 7.2. The optical constants for these calculations were taken from [16].

The maximal reflectivity at the wavelength 50 nm amounts to 17 % (sapphire) and 10 % (fused silica) and strongly decreases for shorter and longer wavelengths, thereby limiting the bandwidth of the OC. Indeed the choice of materials that are transparent at near-1-μm wavelength is large. In [17] over 37 such materials were listed and only 18 materials had data on their refractive index (complex and real parts) for 60-nm wavelength, as tabulated in the book [16]. These materials and their reflectivity for p-polarized radiation at the Brewster angle are summarized in Table 7.1. It is a difficult task to find optical materials fulfilling the above conditions (a)–(f). Materials used in optical applications are usually not sufficiently characterized in the XUV and near-infrared wavelength ranges. To our knowledge, there is no complete data set of material characteristics in the XUV and NIR spectral ranges. The solution

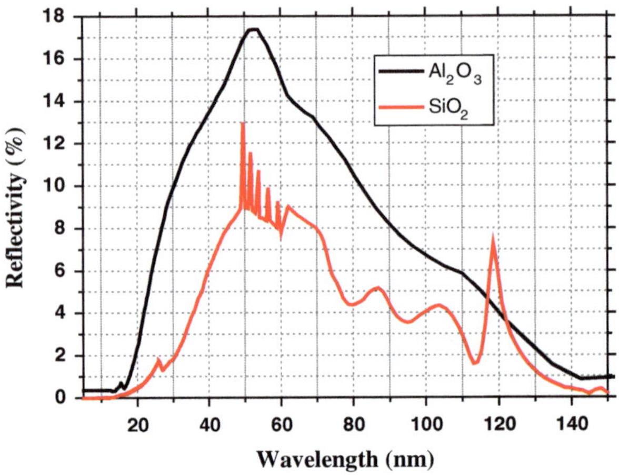

Fig. 7.2 The reflectivity of sapphire (*black curve*) and fused silica (*red curve*) plates for p-polarized XUV radiation at the Brewster angle of incidence for 1-μm radiation [1]

implemented is therefore rather a compromise between the XUV reflectivity and the acceptable optical, thermal and nonlinear properties of the Brewster plate material.

Some of the materials listed in Table 7.1 were experimentally verified in the enhancement cavity. These are MgO, crystalline SiO_2 and fused SiO_2, mono- and polycrystalline diamond,[1] Al_2O_3. The best results with a circulating intracavity power of 12 kW (with no gas jet) were achieved with a fused silica plate due to the low absorption, thermo-optical coefficient and low nonlinearities of this material. However, damage was observed after several minutes. It is necessary to notice that the free-standing SiO_2 Brewster plates were mechanically clamped to the holder thus providing large heat resistance between the plate and holder. Also no water cooling was applied to the plate. More details on the experimental setup and outcome of these experiments can be found in dissertation [18].

One of the main limitations of the Brewster plate method is the low XUV outcoupling efficiency and its relatively narrow bandwidth (conditions (a) and (b)). On the other hand, this method for coupling out XUV is easy to implement.

7.3.2 Wedge on Mirror OC

A thin wedge optically contacted to an HR mirror can be used to separate the DF from the XUV radiation [19]. The device works at the Brewster angle for p-polarized light. The XUV radiation is reflected at the vacuum/wedge interface and the DF propagates through the wedge and becomes reflected by the HR mirror. The small

[1] From Element Six.

Table 7.1 Summary of different IR-transparent materials, and their complex refractive index $n = n_0 + i \cdot k$ at 1 µm and 60 nm, as tabulated in [16].

Material	n_0 (1.03 µm)	k(1.03 µm)	n_0 (60 nm) 20.7 eV	k (60 nm) 20.7 eV	Rp (60 nm), %	Rp (13 nm), %
Si_3N_4 (noncryst)	2		0.617	0.647	29.5	
SiO_2 (fused)	1.45	$10^{-6} - 10^{-5}$	0.862	0.497	7	0.04
SiO_2 (cryst)	1.53		0.79	0.656	12.6	
SiC	2.58	$10^{-4} - 10^{-5}$	0.348	0.481	63	
LiF	1.387	$10^{-7} - 10^{-8}$	1.14	0.73	5	0.015
GaP	3.12	$10^{-7} - 10^{-5}$		0.195		
ZnS (cubic)	2.29	310^{-6}	0.783	0.26	25	0.3
As_2S_3 (vitreous)	2.48	$2.4*10^{-7}$				
InP	3.297	$\sim10^{-4}$			41.5	
GaAs	3.49	$\sim10^{-4}$	1.058	0.162; 0.245; 0.164	12	
C (diamond)	2.39	$10^{-7} - 10^{-6}$	0.5	1.13	46.3	
CdTe	2.849	$\sim10^{-4}$				
NaCl	1.532	$10^{-11} - 10^{-9}$	0.88	0.3	5	
KCl	1.48	$10^{-11} - 10^{-10}$	1.07	0.449	2.4	0.0012
ALON ($9Al_2O_3*5AlN$)	1.776					
$MgAl_2O_4$	1.704					
Al_2O_3	1.755		0.805	0.725	15.6	0.4
Y_2O_3	1.902		1.041	0.542	7	
MgO	1.7229	$\sim10^{-8}$	0.659	0.723	23	
MgF_2	1.37		1.003 (ne=0.894)	0.607 (ne=0.612)	5	
CaF_2	1.428		1.074	0.335	1	
ZnSe	2.489		0.84	0.15	25.6	0.4
KBr	1.544	$10^{-6} - 10^{-9}$	0.97	0.34	3	
CsI	1.756					0.2

Rp (60 nm) and Rp (13 nm) reflectivity for p-polarized light at Brewster angle (for DF) for 60 and 13 nm wavelengths, respectively

angle of the wedge causes separation of XUV and DF. This method has improved the heat conduction properties in comparison with a free-standing Brewster plate. The realization of this method analogously to the thin-disk geometry can make it a truly power-scalable concept. Criteria (a) and (b) are, however, not met by this approach.

7.3.3 Diffraction Grating

An alternative idea consists of an XUV grating etched into the top layer of a highly reflective dielectric coating. The structure acts as a relief grating for XUV light and does not affect the parameters of the DF beam, thus allowing one to avoid any material inside the enhancement cavity and to use only highly reflective optics. The XUV output coupling efficiency in this case is comparable to the Brewster plate method, achieving 10 % for 70 nm wavelength [20]. In a relatively narrow wavelength range, the efficiency can be increased by up to 20 % by fabricating a blazed XUV grating [21]. With this technique, the maximum intra-cavity power level is limited to 7 kW with a gas jet and to 10 kW with no gas jet [8]. The damage mechanism can be related to the localized electric field enhancement of the DF at the grating structure, leading to parasitic losses [21] and UV/XUV irradiation-related issues. The spatially dispersed harmonics may make such an XUV OC unsuitable for generation of attosecond pulses. This method thus does not meet the criteria (a),(b) and (f).

7.3.4 Coupling Through a Hole in a Concave Mirror

Another method is to drill a small hole into a concave mirror right after the focus (mirror *R2* in Fig. 7.1). XUV light has a smaller divergence in comparison with the DF and can thus be coupled out through this hole. The aperture clips the harmonics of lower orders, thus decreasing the XUV bandwidth of outcoupled high harmonics from the long-wavelength side. The hole also introduces losses to the DF and decreases the enhancement factor of the cavity. Very recently, this method was applied to couple out harmonics up to 47th order ($\sim$22 nm) [9] from an enhancement cavity. The power of the coupled out harmonics has yet not been measured.

To couple out longer wavelengths ($>\sim$60 nm) it is therefore beneficial to operate the cavity with a relatively large opening (hole) in the mirror. This is possible at higher-order transverse modes (e.g. TEM_{01}), having minimal DF at the optical axis in the area of the hole. Even then, the higher-order harmonics generated by higher-order transverse fundamental radiation will (a) be of poor efficiency due to the lower intensity in the focus in comparison with the TEM_{00} mode and (b) exhibit a complex spatial profile which may not be applicable for future experiments. So far, this method has shown poor performance in experiment [22].

Another approach called "intracavity mode tailoring" was developed in [23] to avoid these shortcomings. This method uses the degeneracy of several cavity modes to create a beam profile with negligible losses on the opening and reach high peak intensity near the focus (the field maximum is not located in the focus). The distance between mirrors *R1* and *R2* has to be carefully tuned to reach the proper degeneracy. The transverse profile of the generated harmonics and the conversion efficiency are not yet clear for this method.

7.3.5 Non-Collinear HHG

This output coupling method utilizes a completely different HHG scheme. The technique allows generating high harmonics in a direction which is non-collinear with the driving beam [24]. In this case, two circulating DF pulses inside the cavity overlap temporally and spatially in a gas jet. The XUV light generated is directed along the angle bisector of the two driving beams. The process of non-collinear HHG is poorly investigated and may result in conversion efficiencies much lower than those of a standard HHG with a single fundamental beam. Moreover, the output coupling efficiency of this method is limited by the cavity design [24]. Implementation of this method presents a major challenge.

Among the methods proposed for output-coupling the Brewster plate, diffraction grating and very recently coupling through an opening [9] are hitherto the only approaches implemented in enhancement cavities to couple out XUV light.

7.4 Grazing-Incidence Coated Plate (GIP) as an XUV Output Coupler

The idea of an efficient and broadband XUV OC is similar to the Brewster plate output-coupling method: the plate introduces the lowest possible losses for DF and some reflection for XUV. In contrast to the Brewster plate, however AR coating of both sides of the plate is proposed for the grazing incidence of the DF (Fig. 7.3a). At large angle of incidence (>75°), XUV light has drastically increased reflectivity, as follows from the Fresnel equations. This effect can be used to enhance the XUV reflectivity by fabricating a low-loss AR coating for the DF (usually infrared radiation) at both sides of the plate, as shown in Fig. 7.3a. Typically, fused silica is used as low-refractive-index material in multilayer anti-reflection coatings. This material is well studied in both the XUV and near-infrared wavelength ranges. The optical constants of different types of fused silica were measured and tabulated in [16]. Fused silica can be chosen as top-layer-material of the AR coating of the GIP. In Fig. 7.4 , the XUV reflectivity spectra are shown for different grazing incidence angles. At angles of incidence larger than 75° and with s-polarized light, the XUV

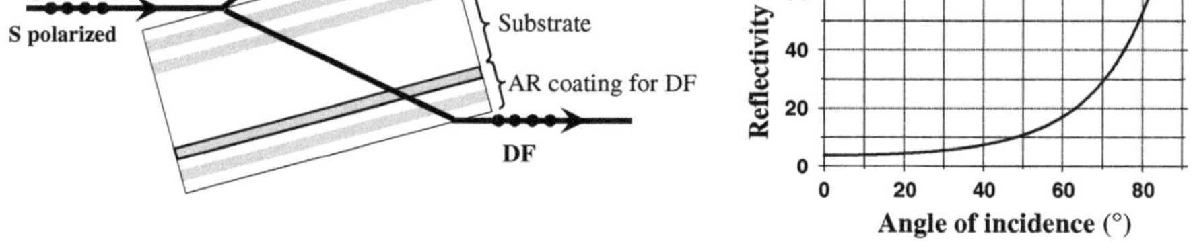

Fig. 7.3 **a** Schematic of the GIP. *HH* high harmonics, *DF* driving field. **b** Reflectivity of the s-polarized 1030-nm wavelength from fused silica at different angles of incidence [1]

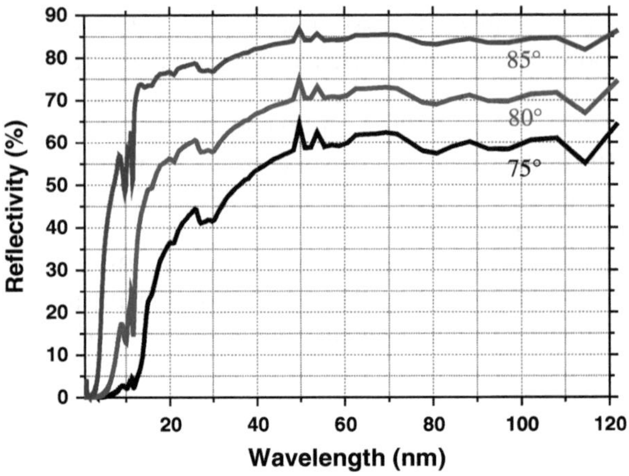

Fig. 7.4 Calculated spectral reflectivity of SiO_2 for s-polarized XUV radiation at angles of incidence 75°, 80°, 85° [1]. The optical constants were taken from [16]

output coupling has reasonable reflectivity (~15 %) even at a wavelength of 15 nm. For the feasible angle of 85°, the reflectivity at 5 nm is expected to be as high as 33 %.

7.4.1 Technical Realization of the GIP

Typically, anti-reflection coatings are designed for normal incidence and suppress only a few per cent of the reflectance (4 % in the case of fused silica) at the material-air interface. In this thesis an AR coating specifically designed for a large angle of incidence is considered. In [25–27] the authors empirically predict a residual reflectance from the AR coating in the cases where total optical thickness and dispersion of the materials and substrates are known. To our knowledge, there is no empirical expression describing the residual reflectance at an angle >75°, and so numerical calculations are needed. In addition, the acceptable angle bandwidth narrows at a large angle of incidence. According to Fig. 7.3b, suppression of the reflectance of 50 % of the incident light is necessary at angles of incidence around 75–80° in order to achieve a low-loss anti-reflective coating. This is the key criterion in fabricating the proposed anti-reflection coating. Another criterion is low sensitivity of the residual reflection of the DF to variations of the incidence angle (the accepted angle of incidence). From the principle of operation, any multilayer coating has an angle-dependent reflection. Due to the divergence of the DF beam incident on the GIP (see Fig. 7.1), the angles of incidence are different for the central part of DF and its periphery. In the case of a typical cavity, this difference can be as high as 1° (The ROC of *R2* and *R1* is 150 mm).

7.4.2 Fabrication and Characterization of the AR Coating

For the GIP presented here, the structure of the anti-reflection coating consists of alternating SiO_2 and Nb_2O_5 layers. These materials have the largest difference of the refractive indices, and so they were chosen as the optimum choice to achieve low residual reflectivity at grazing incidence. The coating was designed[2] with the commercial software package Optilayer for a 75° angle of incidence at the central wavelength 1030 nm. The simulated spectral transmission of the coating is shown in Fig. 7.5 for different angles of incidence. One can see that the losses for the bandwidth-limited sech^2-shaped 100-fs pulse corresponding to an 11-nm bandwidth spectrum are below 0.2 %. It should be pointed out that a robust design of AR coating consisting of only 3 pairs of SiO_2 and Nb_2O_5 layers was chosen in order to reduce the manufacturing time and hence the corresponding accumulated errors. More complex multilayer structures with broader bandwidth and increased acceptance angle of incidence are possible. To cross-check this statement, an AR coating for 87° angle of incidence and s-polarized light was designed with total (calculated) losses below 0.1 %. This design consists of 30 alternating SiO_2 and Nb_2O_5 layers. A magnetron-sputtered (Helios machine, Leybold Optics) AR coating on a high-quality fused silica substrate was produced and experimental measurements of its residual reflection for DF were performed. The transmission spectrum was measured with a spectrophotometer (Perkin Elmer, Lambda 950) at 0° angle of incidence and shows

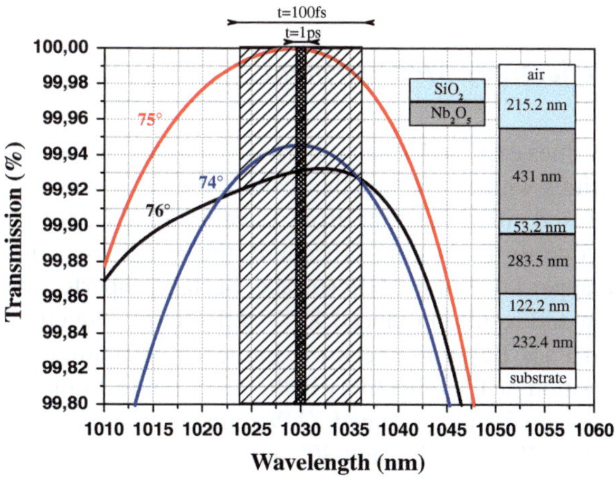

Fig. 7.5 Transmission of the designed AR coating for different angles of incidence. The AR coating was optimized for 75°. The transmission is shown for one side of the two-side coated substrate. The *shaded areas* correspond to the spectral width of 1 ps and 100 fs pulses and show the general tendency of the reduced average reflection for shorter pulses. The *inset* shows a multilayer structure of the AR coating [1]

[2] The design of the AR coatings was done by Vladimir Pervak.

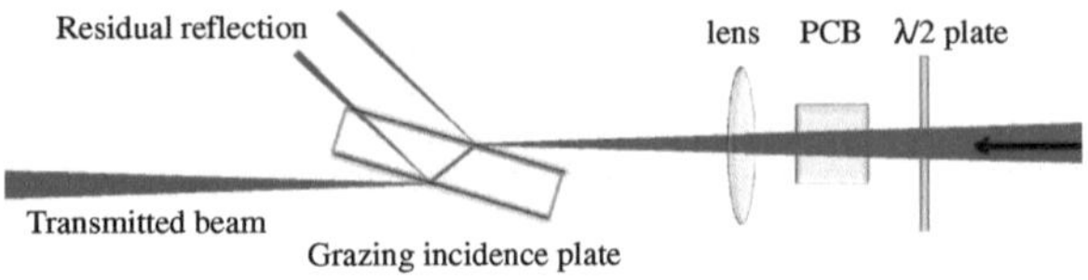

Fig. 7.6 Schematic of the setup for the measurements of residual losses of the GIP

good agreement with the calculated transmission curve. At grazing incidence the laser beam passes a polarizing cubic beam splitter and impinges on the 6-mm-thick GIP in order to achieve a good polarization extinction ratio (see Fig. 7.6). The residual reflected DF power from both sides of the GIP is measured with a power meter and amounts to <20 mW of average power for 10 W of incident power. Therefore, the residual reflection from both sides of the GIP is less than 0.2 % and is in good agreement with the calculations (Fig. 7.5). In order to check the sensitivity of the AR coating to beam divergence, the beam was focused with a 75-mm lens to measure the residual reflection once again. The reflected power remained the same as in the previous test experiment, thus demonstrating the insensitivity of the coating to beam divergence in the range of the radius of curvature chosen for the mirrors. Indeed, the main part of the measured losses was coming from one side of the AR-coated substrate. This side was coated in a second run after filling the deposition chamber with air and rotating the substrate.

7.4.3 XUV Reflectivity Measurements

The reflectivity values expected at such short wavelengths can be reduced because of surface irregularities which approach the scale of the XUV wavelength. The lowest roughness of a fused silica substrate from Layertec GmbH, for instance, is equal to 0.15 nm and, due to the coating procedure, the roughness can be substantially increased at the upper layer of the GIP surface. Surface imperfections cause scattering of XUV light and result in reduced XUV reflectivity. The GIP coating realized was measured at PTB Bessy to check the influence of coating imperfections and chemical composition on the reflectivity in the XUV range. Figure 7.7 shows a comparison of the measured and calculated XUV reflectivity of s-polarized radiation in the range from 1 to 30 nm at 75° and 80° angles of incidence. In the region from 26 to 30 nm the measured reflectivity is around 10 % higher. This may be explained by the different chemical compositions of the materials tabulated in [16]. Additionally, the measured angle reflectivity at 13 nm wavelength shown in Fig. 7.8 is in excellent agreement with calculations based on tabulated optical constants. The difference between the measured reflectivity at 30 nm wavelength and that calculated is pronounced especially at small angles of incidence and vanishes at large angles of incidence.

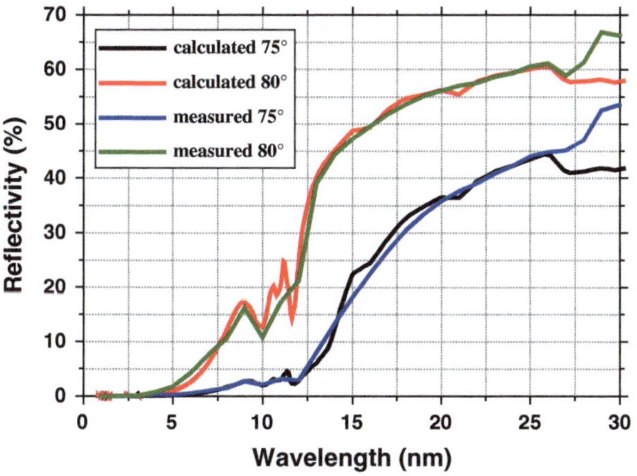

Fig. 7.7 Comparison between the measured XUV reflectivity and that calculated at 75°, 80° angles of incidence [1]

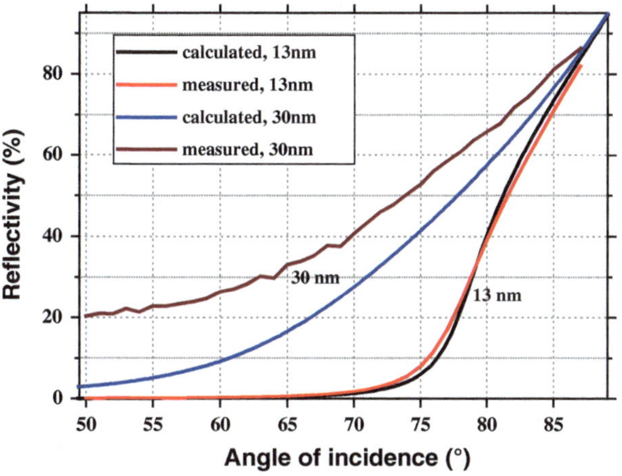

Fig. 7.8 Measured and calculated reflectivity of SiO_2 for s-polarized 13 and 30 nm XUV radiation [1]

The GIP approach allows XUV light to be coupled out in an ultrabroad spectral range. At 80° angle of incidence, the spectral range from 13 nm to at least 120 nm is covered with an efficiency of more than 40 % (see Fig. 7.4). For comparison, the Brewster plate method shows >10 % efficiency in the range 30–80 nm (Fig. 7.2). Due to the large angle of incidence and s-polarized XUV radiation the reflection spectrum of the GIP can be extended even further into the VUV range and beyond.

7.4.4 Limitations and Extension of GIP to Other Spectral Ranges

The implementation of a GIP inside an enhancement cavity (or laser resonator) is straightforward. It can be placed between a focus and a concave mirror in a standard symmetric or asymmetric cavity configuration similar to that shown in Fig. 7.1. An asymmetric cavity configuration (for instance, radius of curvature of the first concave mirror 100 mm and of the second one 200 mm) allows one to extend the distance x between the focus and the concave mirror in order to be able to place the GIP closer to the mirror, thus increasing the spot size on it and decreasing the probability of damage. Such an asymmetric configuration leads to a negligible increase of the beam divergence not influencing the transmission of the GIP. However the phase front distortions induced by the plate can be stronger closer to the concave mirror. Additionally, due to the large angle of incidence, the size of the beam on the GIP (in the plane of the beam in Fig. 7.1) will be even larger. For example, for a 2-mm beam at an angle of 85° the beam size on the GIP will be as large as 23 mm. Technologically, it is a nontrivial task to make it both thin (thus suppressing nonlinear effects) and of high optical quality. However, the technology of coating very thin ($<70\,\mu$m) crystals [28] is routinely used in thin-disk laser technology and allows high optical quality to be achieved.

The damage threshold of the AR coating has to be addressed as one of the main limitations of power scaling with GIP. An AR coating has a higher damage threshold in comparison with a high reflectance coating containing many layers and approaches the damage threshold of the bulk material. This fact has been observed in our labs. The specific value of the AR damage threshold is a subject for further investigation. The grazing angle of incidence increases the effective interaction area at the GIP surface and as a consequence reduces the peak intensity at the sample, leading to a higher damage threshold.

The GIP provides a possibility of controlling the dispersion via the AR coating. A specially designed AR coating can provide noticeable negative group delay dispersion in transmission, similarly to dispersive mirrors [29]. The nonlinear phase of the circulating pulse in the cavity as well as material dispersion can be compensated by a properly designed AR coating. In general, nonlinear effects are a common limitation for all bulk output couplers, which has been investigated in detail [30]. In a GIP, the main contribution to the nonlinear phase is the substrate, due to its thickness in comparison with the AR coating. Not only nonlinear but also thermal effects in a substrate may limit power scaling. The influence of the thermal effects can be suppressed to some extent in a thin and efficiently cooled GIP. Fused silica as a typical substrate material was chosen for the first proof-of-principle realization of a GIP. Other prospective materials such as crystalline quartz, CaF_2 and glasses with low absorption (Suprasil 3002) can be used to reduce thermal lensing and nonlinear phase distortions.

The theoretically designed AR coating has a residual reflection of around 40 % for the p-polarized DF light and has no reflection for the s-polarized light, thus making GIP a polarizer. Polarization sensitivity is a necessary condition for the

Hänsch-Couillaud method of locking an enhancement cavity to the seeding oscillator [31]. The XUV radiation generated polarized parallel to the linearly polarized DF, in our case s-polarized. Delivering the outcoupled XUV beam to the experiment involves further XUV optics, which in general have better reflectivity for s-polarization. In order to realize an enhancement factor of around 100, the residual GIP losses for the DF have to be of the order of 0.05 %. Numerical calculations show that this value is attainable and an advanced GIP can be manufactured with modern coating technologies.

As mentioned above, the reflection spectrum of a GIP can cover VUV, UV and other spectral ranges. The reflection of these spectral components should be considered not only from the upper AR layer of the GIP, as was done for XUV, but from the whole AR multilayer structure. In our specific case, the upper layer of fused silica starts to become transparent at around 150 nm. Above this wavelength interference between reflections from different alternating layers cannot be neglected. Unfortunately, in this specifically described case there is a lack of knowledge about the optical constants of Nb_2O_5 below 400 nm. It is worth noting here that by replacing the alternating material Nb_2O_5 with Ta_2O_5 (the optical constants of this material are known in the spectral range from 150 to 8000 nm) the design shown in Fig. 7.5, has >50 % reflectivity in the ranges 135–140, 142–152, 155–175, 185–215, 240–290, 300–315, 380–480, 500–600, 1200–1700 and 2000–3300 nm, and smooth reflectivity >30 % in the whole range 3500–8000 nm. By varying designs and materials one can expect other broadband and smooth ranges with high reflectivity. New UV-VIS-MIR components can be generated inside the enhancement (or oscillator) cavity by using nonlinear crystals instead of a gas medium.

7.5 GIP for Intra- and Extra-Cavity Experiments

Energy scaling of thin-disk oscillators [32, 33], and fibre and innoslab amplifier laser systems [7, 34] approaches >kW power levels. XUV generation with these systems [35] requires XUV/DF beam splitters able to withstand such high average power. A GIP is well suited to such experiments. Due to the reduced demands on losses of the DF more complex AR coatings with angles of incidence >80° can be realized relatively simply.

Very recently, a proof-of-principle experiment on XUV generation inside an oscillator was realized by Seres et al. [12]. In that experiment a Ti:Sa oscillator with a pulse duration of 17 fs and intracavity power of 10 W was used to reach harmonics down to 30–40 nm. It makes sense to compare this experiment with experiments on XUV generation inside a femtosecond cavity (first and last generation cavities) and intracavity parameters of power-scalable thin-disk oscillators. These data are summarized in Table 7.2. The parameters of these systems are very different and therefore it is difficult to make valid assumptions on the conversion efficiency and XUV output power for the case of XUV generation inside TD oscillators. However from Table 7.2 it is clear that strong focusing down to diameter of 14 μm is necessary

Table 7.2 Summary of different setups for intracavity XUV generation

Setup	fsEc [4]	fsEc [8]	Ti:Sa osc. [12]	TD osc. [33]	TD osc. [37]	TD osc. [this work, [38]]
$\lambda_{driving}$ (nm)	800	1070	790	1030	1034	1030
P_{IC} (W)	38	6500	10	450	1570	310
f_{rep} (MHz)	114	154	9.4	60	60	40
τ (fs)	28	120	17	790	740	200
ω_0 (μm)	5	17.5	<7	**13**	**<7**	**7**
I_p (W/cm^2)	5×10^{13}	7×10^{13}	8×10^{13}	$\mathbf{5\times10^{13}}$	$\mathbf{5\times10^{13}}$	$\mathbf{5\times10^{13}}$

The three left columns correspond to realized systems with XUV generation. No XUV generation was demonstrated for the systems in the three right columns. Hypothetical parameters are 'indicated in bold'. fsEc, femtosecond enhancement cavity; TD osc, thin-disk oscillator; P_{IC}, intracavity average power; ω_0 beam, radius in the focus; I_p peak intensity in the focus

to reach peak intensities of the order of 5×10^{13} W/cm^2 for XUV generation inside TD oscillators. For more details on possible conversion efficiencies and scaling laws see [6, 36]. High intracavity average powers (>1 kW) and pulse shortening due to KLM makes TD oscillators a very attractive alternative to enhancement cavities.

7.6 Summary

The extension of the Brewster plate previously used as an XUV output coupler inside an enhancement cavity is described. The proposed GIP, or grazing-incidence coated plate, has low losses for the fundamental light circulating inside the cavity and serves as a highly efficient, extremely broadband output coupler for XUV. Potentially, the short-wavelength reflectivity of the GIP can reach the keV range. Due to several advantages the GIP concept allows further power/energy increase inside the enhancement (or oscillator) cavity. Further steps becoming feasible with this concept and can cover the VUV-UV spectral ranges and far beyond to mid-infrared. A GIP can be designed for either s- or p-polarization. Owing to its polarization properties, a GIP can be used not only as a dichroic beam splitter but also as a beam combiner or a filter. A GIP is ideally suited to XUV outcoupling and separation inside or outside laser resonators.

References

1. O. Pronin, V. Pervak, E. Fill, J. Rauschenberger, F. Krausz, A. Apolonski, Ultrabroadband efficient intracavity XUV output coupler. Opt. Express **19**(11), 10232–10240 (2011)
2. A. Ashkin, G. Boyd, J. Dziedzic, Resonant optical second harmonic generation and mixing. IEEE J. Quantum Electron. **2**(6), 109–124 (1966)
3. E.O. Potma, C. Evans, X.S. Xie, R.J. Jones, J. Ye, Picosecond-pulse amplification with an external passive optical cavity. Opt. Lett. **28**(19), 1835–1837 (2003)

4. C. Gohle, T.U.M. Herrmann, J. Rauschenberger, R. Holzwarth, H.A. Schuessler, F. Krausz, T.W. Hänsch, A frequency comb in the extreme ultraviolet. Nature **436**, 234–237 (2005)
5. R.J. Jones, K.D. Moll, M.J. Thorpe, J. Ye, Phase-coherent frequency combs in the vacuum ultraviolet via high-harmonic generation inside a femtosecond enhancement cavity. Phys. Rev. Lett. **94**, 193–201 (2005)
6. A.K. Mills, T. Hammond, M.H. Lam, D.J. Jones, XUV frequency combs via femtosecond enhancement cavities, open access paper on arXiv.org, 2012. http://arxiv.org/abs/1206.0408
7. T. Eidam, S. Hanf, E. Seise, T.V. Andersen, T. Gabler, C. Wirth, T. Schreiber, J. Limpert, A. Tünnermann, Femtosecond fiber CPA system emitting 830 W average output power. Opt. Lett. **35**(2), 94–96 (2010)
8. D.C. Yost, A. Cingöz, T.K. Allison, A. Ruehl, M.E. Fermann, I. Hartl, J. Ye, Power optimization of XUV frequency combs for spectroscopy applications. Opt. Express **19**(23), 23483–23493 (2011)
9. S. Holzberger, I. Pupeza, D. Esser, J. Weitenberg, H. Carstens, T. Eidam, P. Russbüldt, J. Limpert, T. Udem, A. Tünnermann, T. Hänsch, F. Krausz, E. Fill, Sub-25 nm high-harmonic generation with a 78-MHz repetition rate enhancement cavity, in *CLEO: QELS-Fundamental Science*, p. QTh5B.7, Optical Society of America (2012)
10. I. Pupeza, T. Eidam, J. Rauschenberger, B. Bernhardt, A. Ozawa, E. Fill, A. Apolonski, T. Udem, J. Limpert, Z.A. Alahmed, A.M. Azzeer, A. Tünnermann, T.W. Hänsch, F. Krausz, Power scaling of a high-repetition-rate enhancement cavity. Opt. Lett. **35**(12), 2052–2054 (2010)
11. J.M. Kaster, I. Pupeza, T. Eidam, C. Jocher, E. Fill, J. Limpert, R. Holzwarth, B. Bernhardt, T. Udem, T.W. Hänsch, A. Tünnermann, F. Krausz, Towards MW average powers in ultrafast high-repetition-rate enhancement cavities, in *High Intensity Lasers and High Field Phenomena*, p. HFB4, Optical Society of America (2011)
12. E. Seres, J. Seres, C. Spielmann, Extreme ultraviolet light source based on intracavity high harmonic generation in a mode locked Ti:sapphire oscillator with 9.4 MHz repetition rate. Opt. Express **20**(6), 6185–6190 (2012)
13. P. Jaegle, *Coherent Sources of XUV Radiation* (Springer, New York, 2006)
14. S. Holzberger, I. Pupeza, J. Kaster, T. Eidam, B. Bernhardt, A. Vernaleken, O. Pronin, V. Pervak, R. Holzwarth, T. Udem, J. Limpert, A. Apolonskiy, E.E. Fill, T.W. Hänsch, A. Tünnermann, F. Krausz, Power scaling limitations for cavity-assisted high-harmonic generation, in *Lasers, Sources, and Related Photonic Devices*, p. AM4A.12, Optical Society of America (2012)
15. B. Bernhardt, A. Ozawa, A. Vernaleken, I. Pupeza, J. Kaster, Y. Kobayashi, R. Holzwarth, E. Fill, F. Krausz, T.W. Hänsch, T. Udem, Vacuum ultraviolet frequency combs generated by a femtosecond enhancement cavity in the visible. Opt. Lett. **37**(4), 503–505 (2012)
16. E. Palik, *Handbook of Optical Constants of Solids* (Academic Press, New York, 1998)
17. D.C. Harris, *Materials for Infrared Windows and Domes* (SPIE Press, Bellinham, 1999)
18. I. Pupeza, Power Scaling of Enhancement Cavities for Nonlinear Optics. PhD thesis, LMU, München (2011)
19. I. Pupeza, E. Fill, F. Krausz, Low-loss VIS/IR-XUV beam splitter for high-power applications. Opt. Express **19**(13), 12108–12118 (2011)
20. D.C. Yost, T.R. Schibli, J. Ye, Efficient output coupling of intracavity high-harmonic generation. Opt. Lett. **33**(10), 1099–1101 (2008)
21. Y.-Y. Yang, F. Süßmann, S. Zherebtsov, I. Pupeza, J. Kaster, D. Lehr, H.-J. Fuchs, E.-B. Kley, E. Fill, X.-M. Duan, Z.-S. Zhao, F. Krausz, S.L. Stebbings, M.F. Kling, Optimization and characterization of a highly-efficient diffraction nanograting for MHz XUV pulses. Opt. Express **19**(3), 1954–1962 (2011)
22. K.D. Moll, R.J. Jones, J. Ye, Output coupling methods for cavity-based high-harmonic generation. Opt. Express **14**(18), 8189–8197 (2006)
23. J. Weitenberg, P. Rußbüldt, T. Eidam, I. Pupeza, Transverse mode tailoring in a quasi-imaging high-finesse femtosecond enhancement cavity. Opt. Express **19**(10), 9551–9561 (2011)
24. A. Ozawa, A. Vernaleken, W. Schneider, I. Gotlibovych, T. Udem, T.W. Hänsch, Non-collinear high harmonic generation: a promising outcoupling method for cavity-assisted XUV generation. Opt. Express **16**(9), 6233–6239 (2008)

25. T.V. Amotchkina, Empirical expression for the minimum residual reflectance of normal- and oblique-incidence antireflection coatings. Appl. Opt. **47**(17), 3109–3113 (2008)
26. J.A. Dobrowolski, A.V. Tikhonravov, M.K. Trubetskov, B.T. Sullivan, P.G. Verly, Optimal single-band normal-incidence antireflection coatings. Appl. Opt. **35**, 644–658 (1996)
27. A.V. Tikhonravov, M.K. Trubetskov, T.V. Amotchkina, J.A. Dobrowolski, Estimation of the average residual reflectance of broadband antireflection coatings. Appl. Opt. **47**, C124–C130 (2008)
28. J. Neuhaus, Passively mode-locked Yb:YAG thin-disk laser with active multipass geometry. PhD thesis, University Konstanz (2009)
29. R. Szipöcs, K. Ferencz, C. Spielmann, F. Krausz, Chirped multilayer coatings for broadband dispersion control in femtosecond lasers. Opt. Lett. **19**(3), 201–203 (1994)
30. K. Moll, R. Jones, J. Ye, Nonlinear dynamics inside femtosecond enhancement cavities. Opt. Express **13**(5), 1672–1678 (2005)
31. T.W. Hansch, Laser frequency stabilization by polarization spectroscopy of a reflecting reference cavity. Opt. Commun. **35**(3), 441–444, (1980)
32. D. Bauer, I. Zawischa, D.H. Sutter, A. Killi, T. Dekorsy, Mode-locked Yb:YAG thin-disk oscillator with 41 μJ pulse energy at 145 W average infrared power and high power frequency conversion. Opt. Express **20**, 9698 (2012)
33. S.V. Marchese, C.R. Baer, A.G. Engqvist, S. Hashimoto, D.J. Maas, M. Golling, T. Südmeyer, U. Keller, Femtosecond thin disk laser oscillator with pulse energy beyond the 10-microjoule level. Opt. Express **16**(9), 6397–6407 (2008)
34. P. Russbueldt, T. Mans, J. Weitenberg, H.D. Hoffmann, R. Poprawe, Compact diode-pumped 1.1 kW Yb:YAG innoslab femtosecond amplifier. Opt. Lett. **35**(24), 4169–4171 (2010)
35. S. Hädrich, M. Krebs, J. Rothhardt, H. Carstens, S. Demmler, J. Limpert, A. Tünnermann, Generation of μW level plateau harmonics at high repetition rate. Opt. Express **19**(20), 19374–19383 (2011)
36. B. Bernhardt, Dual comb spectroscopy. PhD thesis, LMU, München (2011)
37. C.R.E. Baer, C. Kränkel, C.J. Saraceno, O.H. Heckl, M. Golling, R. Peters, K. Petermann, T. Südmeyer, G. Huber, U. Keller, Femtosecond thin-disk laser with 141 W of average power. Opt. Lett. **35**(13), 2302–2304 (2010)
38. O. Pronin, J. Brons, C. Grasse, V. Pervak, G. Boehm, M.-C. Amann, V.L. Kalashnikov, A. Apolonski, F. Krausz, High-power 200 fs Kerr-lens mode-locked Yb:YAG thin-disk oscillator. Opt. Lett. **36**(24), 4746–4748 (2011)

Conclusion

Femtosecond oscillators serve as important instruments in many experiments and are essential elements for amplifier and enhancement cavity systems. The development of such oscillators is of crucial interest. The results presented in Chaps. 2–7 demonstrate reasonable progress in the development of femtosecond high-power lasers. Kerr-lens mode-locked thin-disk Yb:YAG oscillators can lead to many interesting applications in science and technology. Some of which are highlighted in the thesis.

In Chap. 4 an Yb:YAG thin-disk oscillator mode-locked via a SESAM is presented. Many different SESAMs were investigated experimentally, among which only a few samples could be used and all others were damaged due to self-Q-switching. These efforts culminated in the development of an oscillator with 100 W output power, 800 fs pulse duration and 2.5 μJ pulse energy. At this energy level the appearance of CW components in the spectrum was observed. To overcome this limitation an experiment on mode-locking this oscillator in the regime of positive dispersion was carried out, but due to damage to the SESAM no mode-locking could be obtained. The aforementioned experimental efforts lead to the conclusion that SESAMs are limited regarding damage due to the self-Q-switching, surface quality, TPA, heating and the corresponding thermal effects. Simultaneous operation of the oscillator with high power (>1 kW intracavity), high energy (>20 μJ) and short pulses (<300 fs) is limited due to the above-listed shortcomings. Major technological improvements are needed to overcome these obstacles.

Chapter 5 describes the experimental steps in the first realization of KLM in the thin-disk oscillators, motivated by the limitations of SESAM mode- locking. The experimental results of the oscillator operating with 0.5-, 1-, 3-, 6-mm-thick fused silica as Kerr medium are presented. The best performance was reached with the 1-mm-thick Kerr plate and total GDD of $-22000\,\mathrm{fs}^2$. The oscillator was successfully operated with two different output couplers of 5.5 and 14 % transmission. With 5.5 % transmission 200 fs pulse duration and 0.4 μJ pulse energy with 17 W of average power at 15 % optical-to-optical efficiency was achieved. With higher output coupling the parameters were 250 fs, 1.1 μJ, 45 W and 25 % efficiency. These results were obtained with a weak SESAM having a negligible modulation depth $\Delta R < 0.1\,\%$ and

O. Pronin, *Towards a Compact Thin-Disk-Based Femtosecond XUV Source*, Springer Theses, DOI: 10.1007/978-3-319-01511-8,

thermal effects. The chip was developed in the frame of this work. Pure hard-aperture KLM was also realized with similar parameters and a somewhat shorter pulse duration of 190 fs. Peak-to-peak intensity fluctuations of the oscillator are below 0.8 % and are defined by the fluctuations of the pump diodes. The laser shows excellent beam pointing stability with <10 μrad and is insensitive to back reflections. This oscillator configuration was also operated in the positive dispersion regime with the weak SESAM as starter. The average power of 17 W and pulse duration of 1.7 ps with a full spectral width of over 20 nm could be reached at an output-coupling transmission of 5.5 %. The pulses were externally compressed down to 190 fs. These parameters were obtained at nearly an order of magnitude lower GDD in comparison with the negative dispersion regime. It was found in experiment that the oscillator could not be started at the positive GDD exceeding 10000 fs^2.

The Yb:YAG thin-disk oscillator was mode-locked via SESAM and pure KLM in the negative and positive dispersion regimes with practically the same resonator configuration. This allows one to make an adequate comparison of those regimes with each other and with the theoretical model presented in Chap. 2.

In Chap. 6 spectral fibre-broadening and compression of 250-fs, 1.1-μJ pulses from the TD oscillator is presented. The sub-40-fs pulses obtained were limited by the bandwidth of the available chirped compressor mirrors. Spectral broadening with a corresponding Fourier limit of 19 fs was achieved at the full power throughput of 40 W. The compressed output was used to detect the f_{CEO} beat signal, which was stable in amplitude for hours and exhibited fluctuations within a 2 MHz range. Preliminary experiments on the stabilization of f_{CEO} via feedback control of the pump diodes show that eliminating noise from the pump diodes as well as acoustic noise from the active TD cooling should enable CE phase-locking with at least sub-radian stability, making high-field experiments possible.

In Chap. 7 a novel method of coupling XUV light out of the cavity and separating it from the driving field is realized. It consists of a glass substrate having a low-loss anti-reflection coating for the wavelength 1030 nm at grazing incidence of 75° and simultaneously serving as a high reflector for radiation in the range of 1–100 nm with reflectivity >60 %. The device can be used for both extra- and intra-cavity XUV generation experiments.

The power scalability of KLM thin-disk oscillators is supported by the energy scaling laws in the regime of positive or negative dispersion, by the power scalability of the thin-disk concept and the absence of absorption or parasitic intensity-dependent effects in a Kerr medium. The combination of the thin-disk concept and Kerr-lens mode-locking has proved its potential in this work and will very likely be a basis for the next generation of all solid-state diode-pumped thin-disk oscillators.

This work was focused on the development of two main prerequisites for creating a compact XUV source: an oscillator with high intracavity intensity and average power as well as a device able to efficiently couple out the generated XUV radiation. This combination justifies the title of this thesis and should allow compact thin-disk-based XUV sources in the near future.

Appendix A
Summary of Different TD Oscillators

Table A.1 Summary of different thin-disk oscillators with output power above 10 W

Gain medium	Mode locker	Pulse duration (fs)	Average power (W)	Pulse energy (μJ)	Repetition rate (MHz)	Reference
Yb:YAG	SESAM	1120	145	0.941	3.5	[1]
	SESAM	1040	108	30.7	3.5	[2]
	SESAM	928	76	25.9	2.93	[3]
	SESAM	810	60	1.7	34.3	[4]
	SESAM	796	63	5.1	12.3	[5]
	SESAM	791	45	11.3	4	[6]
	SESAM	730	16.2	0.5	34.6	[7]
	SESAM	705	80	1.4	57	[8]
	KLM	270	45	1.1	40	[9]
	KLM	200	17	0.4	40	[9]
Yb:Lu_2O_3	SESAM	738	141	2.4	60	[10]
	SESAM	535	63	0.8	81	[11]
	SESAM	523	24	0.4	65	[12]
	SESAM	370	20.5	0.3	65	[12]
	SESAM	329	40	0.5	81	[11]
Yb:$LuScO_3$	SESAM	235	23	0.3	70	[13]
Yb:$KLu(WO_4)_2$	SESAM	440	21.3	0.6	34.7	[14]
Yb:$KY(WO_4)_2$	SESAM; spectral filtering	240	22	0.9	25	[15]

O. Pronin, *Towards a Compact Thin-Disk-Based Femtosecond XUV Source*,
Springer Theses, DOI: 10.1007/978-3-319-01511-8,

Appendix C
Data Archiving

The experimental raw data used in this thesis, the evaluation files, as well as the LaTeX source code of the thesis can be found on the data archive server of the Max Planck Institute of Quantum Optics, Laboratory for Attosecond Physics, in the following directories:

Oscillator	Master directory, containing the LaTeX code and other important directories:
Chap. 1 Figs, Chap. 2 Figs, Chap. 3 Figs etc.	Directories containing the figures ordered according to their appearance in the thesis
Publications	The journal papers co-authored by O. Pronin, in .pdf format

In the following, a complete list of the directories Chap. 1 Figs, Chap. 2 Figs etc. containing figures, measurement and/or simulation data is given.

O. Pronin, *Towards a Compact Thin-Disk-Based Femtosecond XUV Source*, Springer Theses, DOI: 10.1007/978-3-319-01511-8,

	Directory **Chap. 1 Figs**	
Fig. 2.6	Oscilloscope screenshot	QML.pptx
	Directory **Chap. 2 Figs**	
Fig. 3.3b	Calculated data	CavitySimple.opj Calculated by Jonathan Brons. Standard ABCD matrix formalism
Fig. 3.4	Measurement data	OutputPowerSimple.opj
Fig. 3.5b	Calculated data	CavitySimple.opj Calculated by Jonathan Brons. Standard ABCD matrix formalism
	Directory **Chap. 3 Figs**	
Figs. 4.2, 4.3	Measurement data	SESAMs.opj Measured by Farina Schättiger and Dominik Bauer
Fig. 4.4	Measurement data	absR.opj
Fig. 4.6	Measurement data	Embedded in: surfaceSAM.pptx Measured by Mikhail Larionov
Fig. 4.7	Measurement data	Embedded in: surfaceBatop.pptx Measured by Dominik Bauer
Fig. 4.8	Measurement data	Embedded in: temperatureHD.pptx
Fig. 4.9	Calculated data	cavitySAM.opj Calculated in Winlase software
Fig. 4.12	Measurement data	specsM4838.opj
Fig. 4.13	Measurement data	outputPowerSAM.opj
Fig. 4.14	Measurement data	specsBatop2.opj
Fig. 4.15	Simulation data	EvsGddBatop2.opj Simulated by Vladimir Kalashnikov. Described in text
Fig. 4.16a	Measurement data	chaoticQswitch1.opj
Fig. 4.16b	Simulation data	chaoticQswitch2.opj Simulated by Vladimir Kalashnikov. Described in text
	Directory **Chap. 4 Figs**	
Fig. 5.3	Calculated data	ModeVsd.opj Calculated by Jonathan Brons. Standard ABCD matrix formalism
Fig. 5.5	Measurement data	SpecsSF57.opj
Figs. 5.6, 5.7	Measurement data	specsThickGlass.opj
Fig. 5.8	Measurement data	specs1mm5_5OC.opj
Fig. 5.9	Measurement data	specs1mm14OC.opj
Fig. 5.10	Measurement data	specs1mmHardaperture.opj
Fig. 5.12	Measurement data	noiseKLM.opj
Fig. 5.13	Measurement data	rfKLMSESAM.opj
Fig. 5.16	Measurement data	buildupPDR.opj
Fig. 5.17	Measurement data	spectraPDR.opj
Fig. 5.18	Measurement data	acPDR.opj
Fig. 5.19,	Measurement data	Embedded in: NomarskiAndThermo.pptx
Fig. 5.20		
Fig. 5.21	Measurement data	specs1mmYbLuO.opj

	Directory **Chap. 5 Figs**	
Fig. 6.2a	Measurement data	LMA2535.opj
Fig. 6.4	Measurement data	LMA35compressed.opj
Fig. 6.6	Optical spectrum analyzer screenshot	octavePCForig.jpg
Fig. 6.7	Measurement data	octaveBulk.opj
Fig. 6.9	RF spectrum analyzer screenshot	Embedded in: CEphase.pptx
	Directory **Chap. 6 Figs**	
Fig. 7.2	Calculated data	SapphFS.opj Calculated in Optilayer software
Fig. 7.4	Calculated data	RvsNm85.opj Calculated in Optilayer software
Fig. 7.5	Calculated data	RvsNm85.opj Calculated in Optilayer software by V. Pervak
Fig. 7.7, Fig. 7.8	Measurement data	ARmeasured.opj Measured at PTB Bessy

Curriculum Vitae

Contact information
Name: Oleg Pronin
Address: Ludwig-Maximilians-Universität München (LMU),
Am Coulombwall 1, 85748 Garching
Phone: +49 (0) 89 289 14187
E-mail: oleg.pronin@mpq.mpg.de
http://www.attoworld.de/

Personal data
Date of birth: April 4th, 1985
Place of birth: Ertil, Voronezh District, Russia
Nationality: Russian

Education
2008–2012:
Ph.D. student at Max Planck Institute of Quantum Optics.
Laboratory of Attosecond and High-Field Physics. Group of Prof. Dr. Krausz.
Ph.D. thesis : Towards a compact thin-disk-based femtosecond XUV source.
2002–2008:
Moscow Engineering and Physics Institute (MEPhI, Technical university), Department of Laser physics
Diploma thesis: Optical diagnostics of the plasma lens for a heavy-ion beam focusing system. Supervisor: Dr. Andrey Kuznetsov.
1992–2002:
Elementary and high school

Summer school
08.2006–09.2006:

O. Pronin, *Towards a Compact Thin-Disk-Based Femtosecond XUV Source*,
Springer Theses, DOI: 10.1007/978-3-319-01511-8,

GSI (Gesellschaft für Schwerionenforschung) summer school. Data analysis in Matlab.

Working experience
2012–present:
Postdoc at Ludwig-Maximilians-Universität München.
Laboratory of Attosecond and High-Field Physics. Group of Prof. Dr. Krausz.
2005–2008:
Engineer (part-time) at the Institute of Theoretical and Experimental Physics (ITEP), Moscow.

Scholarships
2008–present:
Scholarship of the International Max Planck Research School (IMPRS) on Advanced Photon Science.
2005–2008:
Honorary scholarship at the Institute of Theoretical and Experimental Physics (reviewed periodically every year)

Language skills
Russian mother-tongue, English, German (basic knowledge)

Sport Tennis, cycling, hiking

Publications

Journal Publications

- O. Pronin, J. Brons, C. Grasse, V. Pervak, G. Boehm, M.-C. Amann, A. Apolonski, V. Kalashnikov, and F. Krausz, "High-power Kerr-lens mode-locked Yb:YAG thin-disk oscillator in the positive dispersion regime," *Opt. Lett.*, vol. 37, no. 17, pp. 3543?3545, 2012.
- V. Pervak, O. Pronin, O. Razskazovskaya, J. Brons, I. B. Angelov, M. K.Trubetskov, A. V. Tikhonravov, and F. Krausz, "High-dispersive mirrors for high power applications," *Opt. Express*, vol. 20, no. 4, pp. 4503?4508, 2012.
- H. Fattahi, C. Y. Teisset, O. Pronin, A. Sugita, R. Graf, V. Pervak, X. Gu, T. Metzger, Z. Major, F. Krausz, and A. Apolonski, "Pump-seed synchronization for MHz repetition rate, high-power optical parametric chirped pulse amplification," *Opt. Express*, vol. 20, no. 9, pp. 9833?9840, 2012.
- O. Pronin, J. Brons, C. Grasse, V. Pervak, G. Boehm, M.-C. Amann, V. L. Kalashnikov, A. Apolonski, and F. Krausz, "High-power 200 fs Kerr-lens mode-locked Yb:YAG thin-disk oscillator," *Opt. Lett.*, vol. 36, no. 24, pp. 4746?4748, 2011.
- O. Pronin, V. Pervak, E. Fill, J. Rauschenberger, F. Krausz, and A. Apolonski, "Ultrabroadband efficient intracavity XUV output coupler," *Opt. Express*, vol. 19, no. 11, pp. 10232–10240, 2011.

Conference Talks

- O. Pronin, V. Kalashnikov, V. Pervak, C. Teisset, M. Larionov, J. Rauschenberger, A. Apolonski, E. Fill, and F. Krausz, "Scalability of mode-locked thin-disk oscillators: issues and scenarios of destabilization," in*CLEO/Europe and EQEC 2011 Conference Digest*, Optical Society of America, paper $CA11_5$, 2011.
- O. Pronin, V. Pervak, E. Fill, J. Rauschenberger, F. Krausz, and A. Apolonskiy, "Efficient ultra-broadband XUV output coupler," in *Lasers, Sources, and Related Photonic Devices*, Optical Society of America, paper AW4A.1, 2012.

Patent Applications

- O. Pronin, A. Apolonskiy, F. Krausz, V. Pervak, "Spatially relaying radiation components," patent application number PCT/EP2012/001541, 2012.
- O. Pronin, F. Krausz, A. Apolonskiy, J. Brons, "Laser device with Kerr effect based mode-locking and operation thereof," patent application number PCT/EP2011/005023, 2011.

O. Pronin, *Towards a Compact Thin-Disk-Based Femtosecond XUV Source*,
Springer Theses, DOI: 10.1007/978-3-319-01511-8,

MIX
Papier aus verantwortungsvollen Quellen
Paper from responsible sources
FSC® C105338

If you have any concerns about our products,
you can contact us on
ProductSafety@springernature.com

In case Publisher is established outside the EU,
the EU authorized representative is:
Springer Nature Customer Service Center GmbH
Europaplatz 3, 69115 Heidelberg, Germany

Printed by Libri Plureos GmbH
in Hamburg, Germany